DEER WATCH

Richard Prior

DEER
WATCH

A Field Guide

New Revised Edition

SWAN·HILL
PRESS

For Charlie, Rory and Minty
who call their farm Roe Hill.
And love watching their deer!

1 • At the Bottom of Your Garden?

Walking is now the most popular outdoor pursuit in Britain. According to the Countryside Agency, no fewer than seven million visitors walk in the British countryside every weekend. We are remarkably fortunate in possessing a massive network of footpaths, bridleways and green lanes taking us quickly into deep country. In addition, the recent grant of the Right to Roam over designated stretches of mountain, moorland and commons widens the scope of our rambles and increases our opportunities for watching wildlife. You don't have to go to the Scottish Highlands or penetrate the thickest woods to see deer of one sort or another. Even close to civilisation small groups can be seen out in the open fields. Look out of the train windows and you may be surprised to see them close to the tracks, not only on open farm land but on the banks of cuttings as you rush through the suburbs. Railways and the routes of old canals have been used to penetrate large urban areas so that many of our parks and gardens have been colonised successfully, especially by the smaller species of deer.

I was having tea with an aged aunt near Romsey in Hampshire when out of the corner of my eye I saw the dainty figure of a roe doe in the

What are they? The first trick is to realise that they are deer – then get out the binoculars. These are roe deer feeding on rape.

garden. The deer came quietly round the shed by the budgies' cage and delicately snipped a bud from one of the treasured rose bushes on the patio. Luckily the old lady did not observe the rape of her roses, nor did she happen to be looking when the doe turned to gaze at us, all eyes and ears, and with the stolen flower drooping from her mouth: an unforgettable sight and one that could be repeated in almost any area of England or Scotland which is not entirely swamped by bricks and concrete. These days they are all around us. Our largest and most beautiful land mammals live literally on our doorsteps. They rely more on people failing to notice them or realise what they are than on being terribly secretive. In our busy lives we don't expect deer so we don't look for them. Because we don't look for them we don't see them! Driving through the New Forest one day with the same old lady we saw twenty fallow deer feeding in a field, at which she blandly remarked that Guernsey cows were splendid milkers!

Charming and fascinating sights await anyone who has an eye for wildlife. The key to seeing deer – anywhere – is interest. Cultivate that interest and the deer are there to be seen. The more you see the more fascinating they become.

So many different wildlife species, from birds of prey to frogs and flowers, are becoming scarcer year by year under the combined pressures of urban sprawl, pollution, chemical sprays and the rest. Deer are the exception. They appear to be able to adapt to changing conditions, not suffering from man's activities, but actually taking advantage of them. The muntjac, for example, is a small Asiatic species which has colonised the whole of the Midlands and may already be our most numerous deer. Once just the final ornament to our stately homes, fallow deer now roam free and are very well able to care for themselves in more than forty-six counties.

This success story is not confined to Britain: roe are spreading into central France and even into Arctic Sweden, while red deer may be found in considerable numbers from Spain to Russia, from Britain to Bulgaria. The reasons for this, like so many ecological questions, are complex and not fully understood but the heartening fact remains that you will never have to search far for deer no matter where you live.

Part of their success and their charm lies in a remarkable ability to adapt to an enormous range of habitats. There are few areas of the country which are too urbanised or too inhospitable to provide a living for one sort of deer or another, or even several. They can live cheek by jowl with people or in remote moorland and mountain areas where man

Look for bunches of hair on barbed wire which marks deer movements. Photo Brian Phipps

is still a rare sight. Their food requirements are wide and opportunistic. Almost anything will do at a pinch – whatever is available. Even some plants that are poisonous to other animals, such as ivy are eaten freely. Yew is another favourite. There was a garden in Dorset where two charming deer came out each evening to clip the hedges: Happily the owner was keener on deer than on topiary.

Britain is unique in Europe in the possession of no fewer than six species of deer: red, fallow, roe, sika, muntjac and Chinese water deer, all living wild in the countryside. Some are native and others introduced, either recently or long ago. Tracing the history of each one in old records or back into the fossils of prehistory is an absorbing hobby in itself, but they are all such staggeringly beautiful animals that most of us want to see them alive and in their natural surroundings.

Where do we begin?

Enthusiasm over deer is infectious. Maybe you have already caught a glimpse of some deer in a park or encountered one by chance, slipping dangerously across the road or starting up from a patch of cover and

A 'Deer Rack' – a well-worn track where they have scrabbled down a steep bank.
Photo Brian Phipps

A muntjac literally at the bottom of somebody's garden. Photo Brian Phipps

standing a moment in alarm before crashing off in wild flight. What sort was it? Do deer exist closer to home? Can you actually get close to one without scaring it away? Look for the answer to one question and you are on the way to solving the next. Don't imagine however that it is all going to be like a television programme, because everything for the small screen has to be condensed. It all looks so easy that one forgets the skill and endless patience of the cameraman. Looking for deer yourself is a real challenge and final success gives a terrific sense of achievement. The first step is to understand the deer's world, so different even though so close to our own.

2 • Deer Lifestyles

To see deer, except by chance, means one must try to understand their world and to see it as far as possible from their viewpoint. About sixty-five million years ago, long after the era of the giant dinosaurs, primitive mammals began to evolve towards their present-day forms. Some became herbivores, primary converters of vegetable abundance, others predators. To be successful, a herbivore needs not only to be able to evade the predator in one way or another, but to have a comparatively high reproductive rate in order to make up for losses. One could say that they are designed to be eaten. As a result the successful deer species breed comparatively rapidly, are well adapted to escape either by hiding or by running away, and have finely developed senses, especially hearing and scent, which are most useful to animals of the forest or at least the forest edge.

Hearing

One only has to see the huge mobile ears of deer to realise how important hearing is to them. They are twin independent receptors which

7

almost always are turning this way and that, even when the deer appears to be nearly asleep, analysing the faintest noises, separating normal woodland sounds from anything out of the ordinary: the squirrel scurrying in the leaves must be distinguished from the approach of another deer; the innocent thump of a fir cone from the predator's stealthy tread. Deer soon recognise man-made noises which by experience they know to be harmless, even the clattering chainsaw and the gas cannons which are put out to protect the crops from pigeons – and deer too! Many forest workers will tell you how tame the deer are, taking little notice of their voices or noisy equipment. Any departure from normal behaviour, however, will produce a very different story. That is the secret of success for the deer which live in close company with us. They are better observers of mankind than we will ever be of deer. They study us and constantly modify their behaviour in the light of that experience.

In remote and open areas like the Scottish Highlands, deer are surprisingly wild, where one would expect the absence of human contact to leave them in a state of primeval innocence. In fact, the rambler will see little of Highland deer unless they are desperate and starving in the snows of winter or the urgencies of the breeding season overcome their timidity. Normally at the distance of a mile or more the sight of a pale anorak moving up the hillside, or a puff of tainted wind will send them off.

Scent

Deer live in a world of scent which is very difficult for us to understand with our blunted senses. If you look closely at a deer's nose you will see that it is hairless, moist and wrinkled. Internally there is a large and complex structure of nasal passages. All this is designed to give deer an extraordinary sensitivity to the most minute traces of scent. This ability is used the whole time. It is a safety check, a means of finding and selecting the most suitable food in a way that no gourmet would ever attempt to match, and a way of communicating with other members of the species, either to warn them away from occupied territory, or to find and join other members of the same group who may have become separated in thick cover.

Deer are equipped with a number of scent glands: on the legs, face and associated with the urinary and genital tract. The way each species uses its powers of scent naturally varies according to their habitat

A sika stag on the alert. All deer are quick to spot movement, but they rely very much on their hearing and acute sense of smell. Photo Brian Phipps

Roebuck browsing. Woodland deer depend for much of their food on the leaves and twigs of trees and bushes. Photo Brian Phipps

preferences and social life. However, you must never forget that no matter how stealthy you are, or how unobtrusively dressed, success in actually seeing undisturbed wild deer will only come if you stay firmly downwind of them. They are well aware of this, of course, and often choose to lie where the twisting, eddying winds of woodland or open hill bring messages of danger successively from one direction and then another. A hint of menace, no matter how faint, will alert them. If they consider the threat sufficiently pressing they will fade into the landscape like last night's dreams.

Vision

Sight in our meaning of the word is less important. An image as perceived by a fallow buck, for example, is probably blurred, lacking definition like a poorly-focused photograph. A human figure is difficult for him to distinguish unless silhouetted, *provided it stays still.* The least movement will be picked up.

The eye of a deer or a human is like a camera, with a lens and iris at the front, and a light-sensitive film, the retina, behind. The main part of the human retina is mostly made up of cone cells which transmit a well-defined colour image to the brain. A deer's retina has few cone cells but many rod cells. These do not transmit colour, so the world to a deer is like a black and white photograph in which colours are transmitted as varying tones of grey. To balance this, the rod cells need less light (the film is 'faster') and they are more sensitive to movement, so although the fallow buck only sees you as a blurred shape he is lightning quick to catch the blink of an eye or the stealthy movement of a hand, even

Listening: Keeping still rather than flight is often the best strategy for a deer in cover.

in deep dusk or thick mist. Anyone who has tried to approach red deer on the hill in Scotland on a day when the cloud is down will know what an advantage the deer have under such conditions. They seem to have X-ray vision through the murk.

It is, incidentally, very rare indeed to find deer completely asleep. They do not seem to need sleep as we do, merely drowsing with eyes half closed but quite clearly alive to the messages conveyed by the wind or through their constantly moving ears. Very occasionally a stalker has come upon a deer stretched out flat which he assumes to be dead, but which has been truly asleep, so deeply as to allow itself to be touched before springing up in sudden alarm. Even in the safety of a park, deer do not generally seem to feel the need for real sleep.

Feeding

Just as deer can rarely afford the risk of going to sleep in case a predator creeps up on them, feeding has to be done in a way that involves the minimum risk. They do this not only by choosing their feeding times when their principal predator, man, is less likely to be about, but by adopting rumination as a means of digestion. Dangerous periods of feeding in the open can be restricted to the minimum time needed to bite off and swallow a reasonable meal of herbage. They then retire into cover or at least to a safer place where they can settle down, head up and senses alert, to chew the cud. Small balls of material are regurgitated from the stomach or *rumen* to be chewed and shredded by the molars into a paste which is re-swallowed, when the whole process is repeated until all the food taken in has been thoroughly mashed up.

What deer is that?

Size	Species	Coat colour (adult) Summer	Spots	Winter	Spots	Antler type in mature male	Tail	Rump patch	Fact sheet Page
Large	Père David	Light brown	No	Greyish	No	Complex. Long back tine	Very long	No	
	Reindeer	Variable	No	Dark brown or greyish	No	Shovel on brow	Short	No	
	Red	Brown or red	No*	Dark brown or greyish	No	Multi-point, round section	Medium	Cream	112–114
	Manchurian sika	Chestnut	Yes	Dark brown	Just visible	8pt round section	Medium	White, black border	128–9
	Fallow Menil	Light brown	Yes	Dark brown	Yes	Palmate	Long	White	60–61
	Common	Brown	Yes	Dark	No	Palmate	Long	White	
	Black	Dark chocolate	No	Near black	No	Palmate	Long	Black	
	White	White or cream	No	White or cream	No	Palmate	Long	White	
Medium	Japanese sika	Chestnut	Yes	Grey or black	No	8pt round	Medium	White, erectile	128–9
	Roe	Red to sandy	No	Dark	No	6pt round	Invisible	White, erectile (winter)	98–9
Small	Chinese water	Red-brown	No	Dull brown, coarse	No		Very short	No	148
	Muntjac	Chestnut	No	Dark brown	No	4pt short with long pedicles	Long	White under tail	87

* Except some park strains

Cud chewing seems to be a relaxing activity for a deer. One sees an animal come into cover from feeding. After standing around for a while and making sure all is quiet it will choose a place to lie down, often scraping with the forefoot before thumping down, first the forelegs and then the hind, so that it is resting comfortably but with all four feet on the ground ready for instant flight in the case of emergency. If nothing happens to alarm the deer, it starts to relax, the eyes half close, the ears adopt a less alert angle, and soon a slight hiccup and ripple up the line of the throat signal the re-arrival of a ball of food. A sideways rhythmic chewing motion follows, like a boy chewing gum.

Hard food may take quite a while to reduce to the right consistency. When it is swallowed again it is not mixed up with the unchewed food but passes to a different part of the stomach, the *omasum,* where the complicated process of digestion continues. Curiously enough, this is not only done by the animal's digestive juices. The most valuable nutrients in a growing plant are the cell contents which are enclosed by stout walls of rather indigestible cellulose. To attack the cellulose, the stomach contains multitudes of primitive organisms, principally bacteria, which break down and feed on cellulose, thus releasing the cell contents for the benefit of the deer. In addition they are capable of producing valuable amino acids from simple nitrogenous sources which pass into the host's bloodstream. These organisms are rather specialised in the type of plant they prefer. Accordingly, the relative concentrations of one or another depend on what the animal habitually eats. For example, a deer feeding on the leaves and twigs of trees will have in its rumen a high concentration of the right sort of organisms for this diet. A change of diet has to be made over a period of possibly ten days or a fortnight to give the rumen flora time to adapt, otherwise there are not enough of the right bugs to digest the new material. Because of this, the kindly notion of putting out hay when deer are hard pressed in a severe winter is unlikely to have much benefit for several days. In fact a deer can starve to death with its rumen full of quite suitable but unfamiliar food.

In normal times cud chewing may go on for quarter of an hour or more. Later the deer may get up and move around, or just loaf about before once more going out to feed. Their preference is not to feed only at dusk and dawn as many people think, but to have short periods of feeding, rumination, rest, and so on throughout the daylight hours and at times during the night when there is enough moon or starlight. Dawn and dusk feeding is only an adaptation for reasons of safety. Completely undisturbed deer will revert to their natural rhythm and may be seen feeding at any time of day. Daylight feeding may also be forced on them in the long summer days when the interval is too long between dawn and dusk, or during a period of dark nights in the winter. In really bad weather deer can become rather torpid, restricting the demand on their energy resources by moving little and thus having less need to eat.

All deer like variety in their diet and they can live on many different foods, from the leaves and twigs of trees to grass, heather, grain, and hedge fruits such as acorns. They are passionately fond of certain fungi. I knew a roebuck who loved bluebell flowers and would walk daintily through the blue carpet snipping individual blossoms and savouring them. The big deer on the whole are grazers, so that with their habit of going about in fairly large groups they can do considerable damage to a farmer's crops. Roe prefer woody browse; coppice growth and brambles being the mainstay, with roses if they can get them! What is a rose but an unusually succulent bramble? The damage they do to young trees brings them into conflict with the forester, although red deer are probably the worst offenders in this respect because they can eat larger trees

Chewing the cud occupies an important part of each day.
Photo the late K. Macarthur

and may, under stress, take to pulling off the bark which can be a very serious form of damage indeed.

Breeding

The young of most deer species are born in early summer. They are very soon able to stagger and then run after mother. For quite a while she does not allow to follow her but hides them while she goes out to feed. They have an instinct to lie doggo and apparently have very little scent. They are startlingly beautiful, wide-eyed and appealing, with a camouflage pattern of white spots or splashes. Don't be tempted to disturb them, or above all to take one away thinking it might have been abandoned. Quite normally a fawn may be left alone for many hours. Imagine the distress and pain of the returning mother who finds her fawn gone, leaving her with a full udder of milk and no offspring to drain it? Yet many people are thoughtless and heartless enough to do this every year.

Deer are, in fact, very good mothers and will defend their young even at risk of their own lives. I have seen a roe doe drive a fox away from her young fawns with vicious blows of her forefeet. A scientist ear-tagging red deer calves told me how he was sent flying by an infuriated mother hind.

Soon the little ones will be following behind or playing like lambs at 'King of the Castle' or 'tag' with other lively youngsters – a sight to watch with pleasure for hours on end in the lazy evenings of a summer deer park.

The six resident species of deer in Britain can be separated into two groups by their behaviour. The large deer, red, fallow and sika, go about in herds, and most of the year the adult males stay in separate bachelor parties from the females and young. The smaller species, roe, muntjac and Chinese water deer, behave rather as we do. Male and female form a somewhat loose association but with fierce loyalty to a chosen piece of ground, the *territory*. The doe will be accompanied by her young of the year, making up the family party, but in contrast to our own, yearlings are unceremoniously thrown out to fend for themselves the following spring when a new family is on the way. One can hear pathetic cries from these astonished teenagers who suddenly find that their mother, for months so attentive, has turned against them and with flattened ears and flailing feet drives them out. Nature in its total

A fallow buck just starting to grow his new antlers.

preoccupation with the species and not the individual can be very unkind.

The rut

The time of year when deer tend to be rather easier to find and when, incidentally, they can do quite a bit of damage, is during the breeding season or *rut*. For red deer, fallow and sika, this comes in the autumn when the big males make themselves obvious by loud challenges, by fighting between themselves, and by thrashing the trees with their antlers near their chosen rutting stands. Red and sika stags roll in malodorous mud made even more potent by their own urine. The scraped bark and broken branches of the trees used as fraying stocks are easy to see – they are intended to be – and help the deer watcher to locate rutting stags. At the same time valuable ornamental trees or young plantations may be ruined. Even the little roebuck does his share of fraying, though the trees he rubs are smaller and a good thistle or a burdock plant often takes the place of something more valuable.

Roe rut earlier than their larger cousins, in late July and early August, but are similarly vocal, barking like small and angry dogs. Muntjac can also be very noisy, but they are non-seasonal breeders because of their sub-tropical origin.

The rut in all species involves quite a lot of physical activity on the part of the male; running after attractive females, defending them from the attentions of other suitors, and apparently bossing the whole affair. In fact, the more we study deer the more clear it becomes that it is not the loud-mouthed smelly male who is in charge of affairs, but the female. She decides where she is going to be for the rut, which male she will accept for a mate, and when she will accept him. Women's Lib has been practised for a million years among red deer! The human equivalent of a rutting stag is the present-day pop singer; both are out to impress the girls. It is up to the females whether or not they are aroused or respond to his performance.

The rut is a particularly exciting time to be out deer-watching and Chapter 4, Deer in Love, covers the subject in detail.

Antlers

Any stag or buck bearing a fully developed set of antlers is a majestic beast, impressive far beyond his size. It is likely that the antlers' function is to impress potential rivals in addition to acting as weapons. Some species, notably the strange-looking Père David's deer, make their appearance even more formidable by rooting up bunches of grass which hang from the stag's already complex antlers during the rut. Once a stag casts his antlers, which he does every year, he loses all his dignity and becomes quite an ordinary-looking animal.

Antlers are quite different from horns. Cows and antelopes grow *horns*, consisting of a hard sheath over a living bony core. They grow throughout life and have a normal blood supply and nervous system. *Antlers*, on the other hand, are solid bone. Early in his first year a male deer starts to grow small bumps on his skull, the *pedicles*. These develop to form the base for the true antlers. Once antler growth starts, elongation is very rapid. Even in deer like the moose which carry immensely heavy antlers growth is still completed in about three months, and the drain on the animal's resources must be severe.

The growing antler at first consists of cartilage which is covered by skin called *velvet* because its short, rather sparse fur has a velvety look and feel. Under this skin there is an abundant blood supply bringing nutrients to the growing point, and the new antlers are rendered sensitive by a network of nerves which are regenerated every year, another unique feature of antlers.

The growing antler is fairly soft and is warm to the touch because of the active blood circulation. Stags are careful to avoid knocking or damaging their antlers at this stage, when quite minor injuries like the prick of a thorn or a small abrasion can produce non-typical growth, sometimes an extra point. A serious impact may half-break the antler and give rise to an untidy malformation. Roebucks in particular seem to be liable to this sort of trouble.

As growth nears completion the underlying structure hardens, the process known as *ossification,* by which the bone is progressively mineralised, the elements concerned being mainly calcium and phosphorus. An overall dietary deficiency of minerals results in the growth of shorter antlers in addition to producing low body-weights, but the deer does not have to take in the extra salts needed while it is actually growing its antlers. Supplies may be drawn from other parts of the skeleton, notably the long bones of the legs. In contrast, the formation of cartilage for the basic structure of the antlers demands an ample supply of protein which cannot be stored to any great extent in the body. Therefore an ample and rich food supply during the period of antler growth is essential for maximum development.

When ossification is complete and the antler is hard, the blood supply is progressively diminished and with its associated blood vessels and nerves the velvet starts to decay. This attracts flies which can sometimes be seen as a black mass on the antlers, and which cause the deer to shake their heads and flick their ears to get rid of them in late summer. They are probably not as much of an irritation as one would imagine because at this stage the antlers are already losing all sensation. Very soon afterwards the velvet starts to peel away from the new antler, which appears white or even slightly blood-stained as the covering is removed. Shreds of velvet dangle from the antlers like rags and a buck or a stag at this stage is said to be *in tatters.* He will assist the process of cleaning off his antlers by briskly rubbing them up and down on any convenient tree or branch, the action known as *fraying.* This is continued after the antlers are cleaned of their velvet until they become polished and stained brown by the tree juices. A buck or a stag using conifers for fraying may finish up with antlers which are coated with turpentine and nearly black.

At this stage the antlers are entirely without feeling except for vibrations transmitted down the beam to the skull. A stag cannot see his own antlers and so he must rely on the experience built up during the time that they were growing and by constant experiment during fraying to use them for fighting or for defence with the precision of a swordsman. When first full-grown they are immensely strong but appear to get progressively more brittle as the year goes on. Fallow and sika in particular may often be seen with broken antlers by the end of the winter.

About six months after fraying the antlers fall off. A line of weakness develops at the top of the pedicle below the *coronet,* the rosette at the base of the true antler. A row of cells becomes decalcified and

eventually the antler topples off or the last attachments are broken by the animal brushing through a fence or jumping down from a bank. Sometimes both fall off almost simultaneously, but at other times some hours or days may elapse between one casting and the other. The deer finds himself unbalanced – a surprising and to him a rather alarming situation. Antler cast may be accompanied by a certain amount of bleeding, but if you look closely at the head of a buck when he is about to cast there is already a swelling round the base which represents new growth waiting to lap over the wound and once more start the process of antler renewal. In a week or so new velvet will already be showing like two mushrooms on the buck's head.

The whole process of antler growth, cleaning and renewal is controlled by hormones in the body under the influence of sunlight. Stags taken to New Zealand, for example, will soon prove this by altering their timetable to the antipodean summer. In simplified terms, antlers grow at a time in the stag's breeding cycle when the male hormone, testosterone, is at a minimum level in the bloodstream. With the approach of the breeding season production of this hormone increases and induces the hardening and fraying of the antler. If an antlered deer is castrated, halting the production of testosterone, the antlers will soon shed and new ones will start to grow, but in a rather uncontrolled way as if some trace of the hormone was necessary for the production of typical antlers. With nothing to bring this to a stop the antlers continue to grow indefinitely. They form what is known as a *peruke*, a shapeless mass of antler. Roe have been known to grow something resembling a wig on the top of their heads, which has given rise to the name.

Once clear of velvet or *clean* (sometimes known as *in hard horn*) the antlers are in constant use for a variety of purposes. The action of fraying is ritualised into an aggressive display to mark the animal's territory. Antlers enhance the animal's physical presence, and are used as a weapon and a shield in offence and defence. Fights between stags or bucks can range from playful fencing between youngsters learning the tricks of the trade to savage encounters between rutting animals which may result in serious injury to one or both of the contestants, even death. The fact that antlers are branched means that it is very difficult for one animal to get through the other's guard without becoming meshed. A pushing match develops, which is unlikely to result in injury to either side. The fallow buck with its flattened palms is very clever in using

In full velvet. Soon the stag will start to fray, rubbing the velvet off to reveal hard antler underneath. Photo Brian Phipps

them to ward off the thrust from the side which is so often more poten-tially dangerous than one from straight ahead. Some animals grow long, straight spikes without points, which can be deadly weapons. A mature roe with unbranched antlers is often known as a *murder buck.*

As there are so many deer in this country you might wonder what happens to the cast antlers. Of course, in parks they are very soon picked up, to make walking sticks, buttons, knife handles and so on. In the wild they are very difficult to find, except by chance, as a freshly shed antler looks very much like a stick. The autumn covers them with leaves and by the following spring they will be bleached on the side exposed to the air and starting to go green with mosses and minute plants. Nature is frugal in her housekeeping and cast antlers are a valuable resource, rich in calcium and phosphorus, which is appreciated by many animals. In the Highlands, where minerals are desperately short, the deer them-selves chew avidly at cast antlers and one may even see a stag's antlers being chewed by another member of his party while he is still wearing them. Find a cast antler that has been several months on the ground and it will almost certainly have been comprehensively nibbled by a variety of small animals. Rabbits, mice, voles and squirrels are all likely to have

Although low in the water, even the small roe is a powerful swimmer.

had a hand in reducing it. There are likely to be the broad toothmarks of the deer themselves and maybe the sharply indented bite of a fox. Even so, it is possible to find cast antlers, and I have a friend in the New Forest who has quite a large collection. Of course he knows where to look and has very sharp eyes.

3 • *How to Find and See Deer*

Making a Start – Are there Deer About?

The first thing to discover is whether there are deer in your chosen bit of countryside. One doesn't need to get up very early in the morning, or actually see any deer. With some notion of how they spend their days, a ramble along local footpaths or woodland rides will soon reveal deer signs. The best way to look for signs of deer activity is to think of yourself as a deer. Look at the woods and fields and hedges and decide where you would lie up for safety, and where you would go to feed. Then go and look to see whether the deer has actually done the sensible thing. Very soon you will start to find evidence of their different activities, not as a surprise and a piece of keen observation but merely as confirmation of something which just has to be there.

Remember as a deer you are not very tall. Even the mighty red stag is only about 105-140cm (41-54in) at the shoulder. Because of their fine bone structure and proud stance deer look taller than they really are, but even a big fallow buck can be entirely hidden in a field of wheat and a muntjac can disappear into grass as if you imagined the whole thing. Their horizon is lower than ours, and the height which they can reach up to feed is also limited. This *browse line* is easiest to see in a deer park, where the skirts of each tree will be cut off as if with shears at a certain height above the ground. If there are many deer in a wood it is equally easy to see, provided you remember to stoop down to the working height of a deer. Doing this can tell you a lot, not only about how many deer are in the wood but, depending on the height, what species they may be. When pressed for food, deer, especially muntjac, will go up on their hind legs to feed, but on the whole they do not bother and the height of the browse line is reasonably reliable as a guide, remembering always that domestic stock, sheep, cattle or goats, may have been there too.

Browsing height	
Red deer	1.5m (5ft)
Fallow or sika	1.4m (4ft 6in)
Roe	1.2m (4ft)
*Muntjac	0.56m–0.86m (1ft 10in–2ft 10in)

*Muntjac frequently browse to the second height by standing on their hind legs. Other deer will do the same, but usually only when pressed for food.

Muddy places are always worth examining for footmarks or slots. Places where deer cross ditches or creep through hedges are called racks. The sliding marks they make as they scramble through are easy to see and will have a lot to tell about the time and direction of travel, the number of deer using the rack, and by their size the sort of deer involved. A word of warning about this: a deer's foot is in two halves, or cleaves, which flex considerably. When the animal is running or sliding the two cleaves separate at the toe, acting as a perfect brake. The mark made is correspondingly bigger and can give a false impression of the size of the animal that made it. The slots of muntjac and Chinese water deer are unbelievably minute. Muntjac, sometimes, but not invariably, have one cleave longer than the other. There is little difference in size of slots between the sexes in the smaller deer, but with red, fallow and sika the difference is quite marked. You will find experts who can tell a great deal from a deer's slots, but take what they say with a pinch of salt. Sometimes they rely on the fact that their expertise can very rarely be challenged! The deer has, after all, walked away. Another point which should not be forgotten is that one or two deer can make an awful lot of slots and one can get a mistaken idea about numbers. An old deer park keeper used to tell me that the easiest way of telling how many deer you'd got was to count the legs and divide by four! It would be a lot more than that if you started counting footmarks.

The droppings of deer are known as *fewmets* if you like to carry on a pleasant medieval tradition, as many people do in the deer world. A healthy deer produces rounded pellets sometimes with a tag at one end. When fresh they are shiny, usually blackish, becoming rougher and brown as they weather. Those of muntjac tend to be faceted, and as muntjac have a habit of defecating in the same place a small mound of fewmets can be discovered in places regularly inhabited by them. The fresh grass of early spring sometimes makes deer scour and a stag in rut is usually rather loose in the bowels.

Bark fraying and stripping

Probably the most noticeable evidence of deer activity is *fraying* where they have removed the bark from bushes and trees. They do this for various reasons: to take the skin (*velvet)* off their antlers when they are fully developed; to mark their territories; as a demonstration of aggression or passion; and possibly to ease the tensions and frustrations of the breeding season (*the rut).*

Fraying height	
(Maximum height to top of damaged bark)	
Red deer	1.8m (5ft 1 1 in)
Fallow or sika	1.6m (5ft 3in)
Roe	0.8m (2ft 7in)
Muntjac	0.5m (1ft 8in)

The height of fraying and the size of the tree attacked give some guide to the type of deer concerned.

Certain species, notably red and fallow deer, may get into the habit of pulling the bark off with their teeth, sometimes but not invariably eating it. This, which is termed *stripping,* may be due to hunger or at least some dietary deficiency, but it also seems be associated with stress. Bark stripping can be distinguished from other forms of tree damage by the marks of the deer's incisors, which make long, near-vertical scores on the trunk.

Deer detective chart

Part damaged	Appearance of damage	Season	Type of damage	Evidence (Width of toothmark)	Height above ground	Species responsible
Stem or trunk	Bark stripped in well defined pieces. Toothmarks parallel and more or less vertical	Dormant	Winter bark stripping	8–9mm	0.3–1.7m	Red
				4–5mm	0.5–0.6m	Roe
					<1m	Fallow/Sika
	Bark stripped in small pieces. Toothmarks parallel and oblique. Often on lower branches and stem	Dormant	Bark nibbling	2.5mm	<0.5m	Rabbit
				3mm	<0.6m	Hare
				<1.3mm	<0.15m	Voles
		Mostly Mar–June	Bark nibbling		Any	Squirrel
	Bark stripped in vertical bands	Active	Summer stripping	Evidence of species present – slots, droppings		Deer
	Bark torn off in strips often on one side. Few or no broken side branches. Wood underneath scored	1 Feb–20 May	Fraying (velvet)		Height <0.8m	Roe
		15 July–15 Sept			<1.8m	Red
		1 July–31 Aug			<1.6m	Fallow/Sika
	Bark torn off in strips, often all round stem, side branches twisted and broken	1 Apr–15 Aug	Fraying (territory)		<0.8m	Roe
		15 Sept–31 Oct	Fraying (velvet and rut)		<1.8m	Red
		15 Sep–31 Oct			<1.6m	Fallow

	Damage	Activity	Season	Measurement	Species
	Bark torn off in strips, often all round stem, side branches twisted and broken	Spring fraying	Mar–Apr	<1.6m	Red
				<1.6m	Fallow
		Fraying (rut)	Nov–Dec	15–38cm / Stems <12mm	Chinese water deer
	Stems prodded into holes (mostly conifers)	Fraying (rut)	Sept–Nov	Stems >10cm	Sika
Buds or twigs	Buds and twigs eaten – cut ends rough with chewed appearance	Winter browsing	Dormant	Maximum height of damage — 1.5m	Red
				1.4m	Fallow
				1.15m	Roe
				0.56–0.86m	*Muntjac
		Summer browsing (mostly hardwoods)	Active	Height-species as winter + indication of species ie tracks, droppings, hair	Deer
	Buds and twigs eaten. Cut ends a clean oblique sheared appearance	Winter browsing	Dormant	Width of incisor 2.5mm / Height of damage <0.5m	Rabbit
				3mm + indications of species ie tracks, droppings, holes / <0.7m	**Hare
	Buds or ends of shoots picked off	Leaf flush	Early spring	On young plants <0.6m + indication of species ie tracks, droppings, feathers	Capercaillie or Blackcock

Note: * Muntjac frequently browse to the second height by standing on their hind legs   ** Bitten twig often lying on the ground below

Four fallow bucks resting in the shade. Go slowly and look carefully or you will miss many deer.

Always remember that deer are not the only animals to eat or damage bark. Careful detective work needs to be done by close examination of any peeled stem before anyone can decide whether the animal responsible was a deer, a vole, a rabbit, a grey squirrel, or perhaps some domestic animal. Quite a lot of damage in fact is done by the forestry workers themselves where thinnings have been pulled out of the standing crop by tractor or winch. There was a serious outbreak of bark stripping years ago in Devonshire and a race of super-squirrels was suspected. Who would have thought that it was actually done by an escaped porcupine! One must be very careful before jumping to conclusions.

Practice is all-important in the game of seeing signs of deer. Once one's eyes are programmed to pick them out, it is amazing how one starts to notice all manner of evidence that most people pass by without a glance. A new and fascinating world slowly opens up. This is indeed our heritage from the wild woods of medieval Britain which, unbelievably, still coexists with the car and the computer, and which is available to anyone with a little time, and patience enough to unlock the door.

Watching wild deer

Some lucky people have wild deer literally on their doorsteps but most of us have to go out and find them. I can remember from my own first steps what a long and discouraging process that can be. I went out day after day scouring the woods, sometimes on hands and knees trying to follow up deer tracks, doing in fact everything wrong and rewarded as a result by little more than the occasional sight of the backside of a deer as it bounced away, or a sound of a distant and angry bark. That is not the way to go about it.

A fraying stock where a male deer (in this case a red stag) has been rubbing the velvet off his new antlers. This often kills the tree and worries the forester.

Ivy is liked by deer and is not poisonous to them. A browse line like this is a good indication of the presence of deer and, judging by the height to which they have reached, the likely species.

Obviously if you are lucky enough to know someone who will take you out deer watching and show you the basic technique, that is the easiest and quickest way to success. One of the best ways of meeting people with real enthusiasm for deer is to join the British Deer Society.[1] Branches up and down the country meet regularly and often arrange visits to parks and other areas of deer interest.

If you are determined to go it alone, do not start at the most difficult end of the business by looking for wild deer, but go about the thing reasonably methodically by spending time studying any captive deer which are within reach of home. There are plenty of deer parks scattered over the country where the public are allowed access. Of course, if you are lucky enough to live in the Scottish Highlands there will be wild red deer within easy reach, and in the late winter they may be as easy to observe as any park deer because semi-starvation will have overcome their natural shyness. Scotland is also fortunate in the number of professional stalkers working full-time on deer management with a long and very fine tradition of knowledge and expertise.

A stalker will not thank you for blundering about the hill in clumsy attempts to watch deer any more than the deer will appreciate being put off their feeding grounds when for most of the year they are balanced on a knife edge of survival for lack of food. So do not bother about your rights to go here or there on the hill, or in the woods for that matter. Discover who owns or looks after the deer, and go and see him. It is very likely that your enthusiasm and willingness to be considerate will be matched by friendly advice about the best places to go and the best times to see deer. Maybe you will be given the privilege of spending an evening in a *high seat* – a deer watching platform placed where there is a good chance of watching wild deer at close quarters. Do not abuse that privilege by leaving the high seat and walking about, or by smoking when the woods are dry and there is a risk of fire.

When you do get out on your own, every hour previously spent watching tame deer will pay its dividend. You will know the colour and size of the deer you expect to see, your eye will almost automatically look for slots in any muddy place, or for the sliding marks which deer have made when crossing a ditch. Signs of browsing will alert you to deer feeding areas and trees or bushes will have been frayed, either in velvet removal, territory-marking, or as evidence of the start of the rut.

[1] British Deer Society, Burgate Manor, Fordingbridge Hants SP6 1EF Tel. 01425 655434

This is the master class of the deer watching business. The techniques of real stalking have to be learnt and practised before there is any real chance of consistent success. The direction of the wind is, of course, paramount, so no matter what the weather forecast predicted the night before, stop the car before you get up to your stalking ground and make sure of the general direction. Holding up a wetted finger is the usual method. It will feel colder on the windward side. In very calm weather a puff of smoke from lighting a match is enough to betray the general drift. Some stalkers even have a child's bubble kit and from time to time send a stream of bubbles into the air as they go through the woods, checking for swirls and eddies which so often give one away to the deer.

The next thing to study is the art of moving slowly enough. It is very difficult to convince yourself that there is just as good a chance of seeing deer where you are as round the next corner or in the far clearing. The pace must be slow enough to give your eyes time to take in the minutest sign of the presence of deer: a vague movement in the bushes, a line of a back or the silhouette of a pair of ears. Patches of colour which are out of keeping with the woodland scene must be examined minutely. Nine times out of ten a russet patch will be dead bracken or a damaged tree turning brown. The tenth time it will be part of a deer which will spring to sight once your attention has been focused on it. With roe in particular, their white targets are a giveaway in the winter. Any suitably shaped patch should be studied and all the possibilities must be eliminated before you move another step.

The other reason for going terribly slowly is to avoid making the least unnecessary noise. The sudden crack of a small twig in the silence of the dawn woods is enough to alert every deer within hundreds of yards and, of course, any metallic noise is much worse. I used to go stalking with a man who had the nervous habit of jangling his car keys in his pocket. It pays to leave small change behind and to put the car keys in a hip pocket or somewhere where they won't rattle.

Clothing and equipment

The best clothes for deer watching should not only tone in with the colours of the woods or the hill, but should be soft in finish so that there is no constant rustling and swishing as you move. Brambles or the spiny branches of Sitka spruce can set up a noise like brushing a live microphone as they drag over the surface of an average waterproof. To begin with at least, an old tweed jacket or a heavy sweater and warm trousers in an unobtrusive colour will do very well. It is tempting to take an anorak but most of them have the dual disadvantage of bad colours and material which rustles. Anything which shines should be avoided and this cuts out a lot of the cheaper waterproofs. Even so, getting wet is nobody's fun and especially in the mountains everyone has to take precautions against sudden changes in the weather and the ensuing risks of getting cold or snowbound, or lost.

The answer is to take a few supplementary items with you in a bag, preferably one that fits over both shoulders to leave your arms free. In this goes a spare sweater, some sort of waterproof, and the survival aids: knife, compass, whistle, first-aid kit, matches, clean handkerchief, anti-midge cream and a piece of string. Any items that must be kept dry can be stored in small polythene bags which with the clean handkerchief can make an effective dressing for a wound which is too large for first-aid plasters. I like to take a map with me because one can learn the lie

Deer droppings, or fewmets, *are a good indication of the presence of deer.*
Photo Brian Phipps

of the ground quicker by referring to it as one turns and twists. When the cloud comes down on the hill the map can, of course, be a life saver, used in conjunction with the compass. A small notebook and pencil slip into a pocket.

Footwear has to be comfortable, otherwise walking is a misery, but must also be chosen according to the ground and time of year. For summer stalking in dry weather a pair of trainers will do very well for the evening. One can feel the ground as the foot is put down and thus avoid treading on a lurking stick. Their light colour does not matter very much because most of the time your feet will be hidden in the grass. By morning, however, there is likely to be a dew and after quarter of an hour of steady progress canvas shoes will be saturated and you will squelch loudly at every step. Apart from the discomfort, the noise will be too much of a give-away and something more waterproof will have to be chosen, even if it is less than ideal from other points of view. Ordinary wellies are hopeless. They are clumsy and noisy. The thin ones that fit tightly up the leg are much better, but still difficult to walk in with sufficient stealth for real close-range woodland work. Talk to six stalkers and they will probably give you six different answers, but on the whole the thinner and lighter one's shoes for woodland work, the better.

On the open hill it is different. For one thing, stalking red deer is a matter of long-range work. Silence is less important. In fact, all one's efforts should be towards observing deer at leisure from a distance rather than creeping close and spooking them. The distances may be long and the going rough, so boots are indicated to give ankle support, with patterned soles for grip. Many professional stalkers still carry on the tradition of wearing heavy nailed hoots. They certainly grip better on slick, wet rocks, but the beginner would be better off with Vibram rubber soles which grip well without being noisy. Leather boots are much kinder on the feet than rubber, but no matter how carefully one treats them with waterproofing, leather boots are usually sodden by the end of the day on the hill in Scotland where dry going is the exception. One can now get lightweight lace-up ankle boots laminated, to make them waterproof with a good secure pattern on the sole without being so thick that one cannot feel for a betraying stone or stick.

Binoculars

There is one expensive piece of equipment which one has to buy to get much pleasure at all out of deer watching, and that is a good pair of binoculars. You need them to spot deer at a distance and to see what they are doing without going too close; you need them in the woods to unravel the complex three-dimensional view of tree trunks and leaves and twigs, into which the deer fade like the giraffe and the zebra in Kipling's story about how the leopard got his spots. 'One – two – three! And where's your breakfast?' Having at last found a deer you will want to feast your eyes on him, study his expression and the shape of his antlers, or try and make out what he is eating, all this usually in poor light, and with speed the very essence.

Many people make the mistake of buying high-powered binoculars. Magnification not only increases the apparent size of the object in view, it also magnifies any unsteadiness on the part of the user so that anything over 10x really needs some form of support, which is rarely available in deer stalking. The field of view is reduced and so is the depth of focus so that half the time you are fiddling with the focusing ring trying to get a sharp image, instead of getting on with watching the deer. High power is also more demanding on the quality of the lenses so that price for price you get much better value if you choose a model with low magnification, say, 7 or 8x.

Small lenses cut down the amount of light transmitted. The pupil of the human eye is capable of dilating in poor light to a diameter of 7mm and to match this a pair of binoculars for use at dusk and dawn, when so many interesting things are going on in the deer world, should have an exit pupil which is not smaller than 7mm. Binoculars are usually marked with two figures: the magnification and the diameter of the objective (the lens farthest from the eye). The exit pupil is found by dividing the objective diameter by the magnification. For example, the exit pupil of a pair of 8 x 30 binoculars is 3.75 but in a pair of 7 x 50 binoculars, a very popular size among deer stalkers, works out at 7.14. A binocular of 8 x has to have objective lenses at least 56mm in diameter to achieve the same light transmission. Glasses of that size are heavy, bulky and expensive.

Whatever binoculars you choose, they should be hung round your neck with a wide strap ready for instant use, not kept in a box as if you were going to the races. A flap of leather hung over the eye pieces keeps

out the rain better than the lens caps which are sometimes provided.

If you get really involved in deer watching there are catalogues full of tempting items of stalking equipment, some of which are very handy, but the basic kit need not be expensive. Like anglers, the longer you have been at it, the less you tend to carry around with you. If I am going out deer stalking in the summer woods I usually wear a lightweight showerproof jacket, a tweed hat, nondescript trousers and rubber-soled leather shoes or rubber boots according to the going. I have a rucksack with spare oddments and a waterproof, and a pair of low-powered compact binoculars round my neck. I take gloves or mittens because a flash of pale hands as you raise the binoculars may give you away. I have a long hazel staff as high as my forehead which is of great assist-ance in steadying the binoculars and yourself on difficult ground. I never attempt to keep my legs dry. Walking in anything waterproof is cramping and noisy.

For watching deer in the hills I take a waterproof jacket that opens fully down the front for ventilation when it gets hot going uphill. Breeks and stockings plus canvas gaiters take the place of trousers as they are better for crawling. A telescope in addition to the binoculars gives added pleasure for detailed deer study.

There are some tempting camouflage garments available which certainly help you to stay unobtrusive and have a definite place among specialists. These days a member of the public seeing anybody in full cammo gear may leap to the wrong conclusion and phone the police. Normally drab clothing is sufficient and certainly less likely to cause alarm.

Tackling an unknown wood

Once permission to watch deer has been obtained how does one go about actually finding them in an unknown stretch of woodland? The first thing to do is some leg work; go to the farmers round about and ask them what they see, explaining your interest. Walk the woods during the day taking care to behave as far as possible like an ordinary tripper, that is not skulking about but walking purposefully and noisily here and there, and keeping, wherever possible, to the rides and tracks, not penetrating into the thicket. Nothing frightens or worries deer more than a human who steps outside a normal pattern of behaviour, and the intention of this exploration is not primarily to see deer. Stroll round the wood, take somebody with you if you like, and talk to them, but keep your eye peeled all the time for signs of occupation. Has the ivy been nibbled off the trees? Are there well-used tracks leading out to the fields? As you go, note likely feeding places; clearings in the woods, wide rides with browsed bushes on either side, vantage points from which you might be able to get a wider view.

If the wood happens to be shot regularly for pheasants make a point of meeting the gamekeeper and offer to join the beating team on shooting days if you possibly can. It will get you to corners of the wood you might never otherwise penetrate, and you may see bedding scrapes or accumulations of droppings which give you an idea of the most used areas where the deer are likely to be during the day. Friendship with the keeper and a willingness to be useful are almost sure to be repaid.

After this preliminary surveying, turn up at first light and look in as many of the surrounding fields as you can to try and pick up deer grazing. This can probably be done conveniently from by-roads without either disturbing the deer or trespassing. Then, as the light strengthens and the deer start to think about lying up for the day, choose a downwind entry to the woods and go slowly round to one or two of the likely places you found in daylight. Each opening or space under the trees must be meticulously spied, first with the naked eye and then again with binoculars. There is usually quite a lot of movement after the sun gets up, especially after a cold night. Your patience may be rewarded by seeing a doe and fawn at their morning games, racing round in the sunlight obviously glad to be warm once again. The grazing deer will be drifting in from the fields, often standing about more or less motionless before bedding down to chew the cud. Roe and muntjac may be on their feet to snip a few extra leaves before the start of the workaday world sends them into cover.

In the evening, deer on the whole will be more vigilant and they are unlikely to be on the move much before dusk. Even in hot weather they do not troop down to the nearest stream to drink like a herd of zebra. Although deer are quite at home in the water and swim much more strongly than one might imagine, they have little need to drink. In our climate there is usually enough moisture on the herbage, rain or dew, to keep them going. It does, however, affect their feeding habits. On summer evenings deer are unlikely to be active until the dew starts to fall and then one should look for them in the valley bottoms and other quiet places which reach dewpoint first.

In your enthusiasm do not stalk the same piece of woodland too often, or at the same time of day. With your sneaking, unobtrusive habits you will be an object of intense suspicion, even though your intentions towards the deer are entirely benevolent. They will, of course, be aware of you, the time you arrive, the route you take, and the time you go; for

A deer footmark or slot. Look for narrow, pointed impressions, though if the deer is moving fast, the two halves (cleaves) of the hoof divide. Sometimes you can see marks of the dewclaws behind.
Photo Brian Phipps.

even though you take care, as you must, always to stalk upwind or at least across it, you leave behind a wash of scent which will declare your excursion to every deer in the wood. After as few as two or three excursions over the same route, the number of deer you see is likely to drop. They are not fools. A man with a chain-saw is soon identified as harmless. Even the gamekeeper who has his regular rounds carries out his duties without being particularly quiet. It is the unexpected and powerful whiff of human scent, or the crack of a stick when there has been no warning of anyone's approach which alarms and disturbs them.

Deer and the weather

In some ways, deer are very human in their reactions to the weather, but dry windy weather makes them nervous because they can no longer trust their noses to the same extent. Rain driven by a brisk wind will send them into the shelter of the hedges where there is a lee, but they don't like drips. Sometimes in really heavy rain there will suddenly be a great many deer to be seen as they change their ground. Immediately the sun comes out after a shower they will be tempted out into clearings and on to the fields to dry their coats and feed.

A sika reaches up to browse on twigs. The height to which they can reach indicates the likely species. Heavy browsing creates a browse line *which can be seen here.*
Photo Brian Phipps

Snow and dry cold in the winter is less of a hardship to them than one might imagine. Just before a snowstorm there may be a period of intensive feeding, but once it starts they prefer to bed down and let it bank up round them. I have seen roe deer curled up like dogs in deep snow, while the bitter wind went over their heads. Under these conditions they feed sparingly and lose very little condition providing they do not have to move or struggle in the deep drifts. The testing time for all deer comes at the end of winter when their reserves are low and their winter coats, once shiny and plump like an eiderdown, have become ragged and lack-lustre. Particularly in the Highlands, a period of sleet or cold rain in March takes a terrible toll of the red deer. Even in the south I have seen deer so chilled that they were literally shivering when a sleet storm coincided with their change from winter to summer coat.

One tip in trying to interpret the effect of weather on deer is to look at the behaviour of any farm stock out in the fields on the way to your stalking grounds. If the cows are all under the thickest hedge, expect the deer to be in shelter. If they are moving freely over the fields munching away, deer will probably be feeding too. Do not, however, expect deer to be feeding in the same field with farm stock. They will tolerate cows to some extent, but a flock of sheep will drive the deer right off and even though they were feeding in it regularly they are unlikely to come back to a field for several weeks after the sheep have gone again.

Talking to deer

You can talk to deer in your own language, or attempt to do it in theirs. The very fact of talking out loud reassures deer that you mean them no harm. They are accustomed to passers-by talking, even shouting, providing they keep to normal human paths, and deer in regular contact with humanity will take little notice. That is not to say that loud conversation is the key to success in deer stalking, but voices approaching will be the signal for a delicate withdrawal into cover rather than panic flight.

Sometimes skilled stalkers will come face to face with a deer at very close quarters. If surprise is complete the deer will be unable to decide what to do and will stand still, racked with spasms of apprehension. There have been occasions when starting to talk quietly to the deer has had a calming effect. I remember vividly encountering a roe doe who was no more than four or five yards from me, whose shivering stopped under my blandishments. After two or three minutes she relaxed

Deer soon recognise farm activities which are harmless to them, like these roe, but stop to look at them and they are off.

completely and took no further notice of me. Of course this will not happen every time and panic flight is the more usual outcome.

One can also imitate various noises which deer make in order to attract them. The most dramatic and exciting is to attempt to call a red stag to you in the rut by imitating the roars of a challenger. It is a delicate business to pitch your voice so that the challenge is neither too menacing on one hand, nor too juvenile on the other, so that the supposed rival is not worth seeing off. If your roar is answered by the other, especially in the gloom of the forest, be sure that there is a good stout tree to hide behind! Mistaken identity can go dangerously far. Fallow bucks do not respond well to a challenge for some reason, although a buck will sometimes show himself if you are successful in imitating his loud, rather pig-like noise.

Roe too are difficult to lure by a male challenge. The language of their barks is varied and subtle so you must understand what to say, otherwise you will find that you are giving not the challenge bark but the alarm call, and that gets you nowhere. It is female squeaks which are the downfall of roebucks who may be lured during the rut from several hundred yards away by the piping of a doe in season. This is a technique which has been developed for many years in Central Europe and has now many competent practitioners in Britain.

It is probable that the other deer species can equally be led astray by encouraging female noises, but this is a field which so far has been very little explored by deer enthusiasts. Like so many other areas on the map of deer behaviour, this is still marked Unknown Territory. How reassuring it is for the rising generation of deer watchers that this should be so.

4 · *Deer in Love*

The breeding time or *rut* of several of our deer species is such an exciting and dramatic time to watch them that explaining what is going on is worth a special chapter. To be out just at the right moment to catch the height of the rut and with the best chances of seeing rutting activity needs careful forward planning. There is, of course, quite a lot of local and seasonal variation in the date one hears the first fallow groan or the first sika whistle, just as there is in the arrival of the first cuckoo from year to year. However, the table below will give a good indication of when to expect action.

Rutting times		
Species	**Approximate duration**	**Expected peak**
Red	late September to late October	10 October
Fallow	early October to late October	20 October
Sika	early October to early November	20 October
Roe	mid July to mid August	7 August
Muntjac	not seasonal	
Chinese water deer	December	
Père David's deer	June to July	

A red stag roars his challenge surrounded by his harem. Photo Brian Phipps

Because the deer species are so different in size and appearance, most of us probably start by thinking that each will have its own rutting behaviour without too much in common with any of the others. This is true to the extent that each species has evolved a set of signals between male and female which are so finely tuned that they are meaningless to another species. In this way the problem of mis-mating is avoided. One can see this best in a park containing more than one species where perhaps red deer and fallow are both in full rut but taking not the slightest notice of one another. It is rather like two conversations going on in a restaurant, one in English and one in Chinese. To each party the other's speech is a meaningless gabble.

The signals by which members of a single species recognise one another must primarily concern their scent; everyone is fully aware these days of the subtle and powerful effect of pheromones. Otherwise the signals can also be audible – the various grunts, groans, squeaks and whistles which are used by both sexes – or visual – in their appearance and behaviour.

Even so, there is a great similarity between the deer species. The more one studies them the more it becomes clear that they do share a common heritage, an approach to life if you wish. A great deal may be learnt about one kind of deer by some form of parallel behaviour in another which may happen to be more obvious. For example the way sika stags have of positioning themselves during the rut on the paths which hinds use between feeding and lying up may seem very different from typical fallow rutting behaviour, where one sees a buck on his rutting stand parading round his does and groaning repeatedly in challenge. It is entirely up to the sika hinds whether they come along that path or not. From this it is not too big a flight of imagination to start speculating what it is that keeps the fallow does in close proximity to the buck. Is it male domination or are they there because they fancy him more than another male? Seeking answers to such questions is vital unless we are just going to enjoy the rut as an animal spectacle without any attempt to enlarge our understanding of their world.

The first illusion which must be dispelled is that any deer community is male-dominated. Poor chaps, with their expensive antlers, smelly habits and loud voices, they are only out to impress the ladies by every means in their power. There is, of course, competition between males, but in the end if a female will not accept a particular individual there is nothing much he can do about it. For all their studied air of indifference, even disdain, female deer from the largest elk to the smallest water deer will stay in the company of a male purely because they fancy him and are willing, maybe, to mate with him. In terms of survival, a male is of entirely secondary importance in a deer community.

Scent

Even to our rudimentary senses a rutting stag smells pretty strong, while most deer farmers can detect when a hind is in season by subtle changes in her odour. The impact on a deer's sensitive nose can be imagined. Watch a dog sniffing a lamp-post and it is quite clear from his behaviour that he is receiving a wide variety of messages and information about previous passers-by. The language of scent is practically unknown to us. To make up, to some extent, for our lack of nose we have to make what deductions we can from meticulous observation of behaviour under as many differing conditions as possible.

Hormonal changes in the blood trigger a chain of responses throughout the body which culminate in the urge and the ability to mate. So far as the male is concerned, the process starts with the growth of antlers which commences in the comparative absence of male hormone. This is needed later to harden and clean them of velvet and turn them from mossy outgrowths into polished fighting weapons which at the same time increase the animal's apparent size. Experiments with caribou in which an individual bull was fitted with detachable antlers of different sizes showed clearly how fundamental the size of an animal's antlers is to his standing among his fellows. The effect of large, impressive antlers on the opposite sex is something which has not been studied enough. Look again at the park situation: why has one stag managed to

Serious fights can develop between dominant males of the herding species, but usually a pushing match decides the master. Photo Brian Phipps

attract forty hinds round him while another has only six? The male domination school will say that the more successful stag is better at rounding up his hinds. Spend an afternoon in the park quietly watching and you will see how far from the truth this really is.

In the run up to the rut various glands increase in size and activity. The testes gain weight rapidly and sperm production is commenced. At the same time accessory glands are increasing their activity. Urine is used as a vehicle for the transmission of scents which are very significant in reproductive behaviour. It may be used to saturate the ground of a scrape or wallow in which the animal will then roll, covering himself with mud which to us smells extremely foul. He may also spray urine directly on to his legs and the underside of his body.

In red deer, fallow and sika, the pocket of skin just below the eye known as the suborbital gland increases in size, and may fill with a brown waxy secretion. There is also a glandular area on the forehead of most deer which increases in activity towards the rut, scent being wiped from it on to any convenient vegetation. Sika stags during the rut may produce a milky fluid from the corner of their eyes. This is peculiar to the species and has not been properly investigated. The muntjac has an additional pair of glands between the eyes and both these and the suborbital glands are capable of gaping open at times of emotion and also when the animal is defecating. The suborbital glands on red, fallow, and sika also gape open during intense rutting activity. It will be noticed with the larger deer species that the penis sheath changes in shape during the rut. In fact the last part of the sheath becomes everted, which exposes a number of scent glands. The large size of these glands suggests that they contribute to the strong odour which is characteristic of a rutting male deer.

The female contribution is less flamboyant. Although the physiological preparations in the male for the breeding season have been in progress for a long time under the influence of sunlight acting on the pituitary gland, it seems likely that the final ferment of the rut is only triggered by the scent given off by a female in season.

Voice

Loud calls are characteristic of most of the herding deer species and very exciting they are to hear in the bare echoing corries of the Scottish Highlands or the dim recesses of the forest. While much deer behaviour seems similar to us, these calls are startlingly different between species. Having heard the guttural roar of a red stag, the bugle of a wapiti and the triple whistle of a sika, it is difficult to remember that these three species are so closely related, in fact, that they can interbreed. One assumes that the calls may be the means of identifying a species at a distance. Hearing them, other males will respond either by returning the challenge or by making themselves scarce if they feel themselves outclassed. Thus a primary purpose has been established by the dominant male, that is of avoiding the risk of having to fight in order to maintain his position. He also makes his appearance as formidable as possible and intimidates rivals by threat displays.

Rutting cosmetics

As we have seen, the larger deer species anoint themselves with scent; red and sika adding to the general effect by stirring up a mixture of mud and urine with their hooves and then wallowing in it. This not only makes them smell but they emerge from the wallow dark with a plaster of mud. If you see two stags on a Scottish hillside, one in the brown coat of late summer and the other black and peaty from his wallow, even though they are identical in size the black one will look bigger and more formidable, and well he knows it. He is more impressive to an adversary and maybe to the hinds as well – who knows? This warpaint serves to reinforce an enhanced appearance of size which was commenced by the growth of large spreading antlers and, just previous to the rut, the development of a swollen neck and a dark, rough mane.

To our eyes the most eccentric of these cosmetics is the habit of Père David's stags of gathering up large tufts of dry grass on their antlers until they look like moving haystacks. These deer also use the long rear tine on their antlers as a scoop to spoon mud on to their backs in order to improve the general effect.

Aggressive displays

For most of the year, the males of the herding species, that is, red, fallow and sika, congregate in bachelor parties and live together in reasonable harmony. Towards the rut they become restless and quarrelsome and as their antlers harden and clean, tempers clearly run short. The groups break up, individuals going off in pursuit of love. Having set up his rutting stand, the master stag or buck has to fend off a constant flow of competitors which he will first warn by his voice and scent signals. If

A pair of roe in the rut. Sometimes these chases produce a marked path, called a ring.
Photo Justin Cowtan

these are not sufficient the next stage is a display of strength and aggression which usually takes the form either of fraying the antlers on the nearest tree, or failing that, of rooting in the heather or grass and sending up tufts of vegetation with the aid of the antlers or a flailing forefoot. One can see a bull do the same thing, scraping with his forefoot or kneeling down to dig furiously in the earth with his horns. This display can be guaranteed to have a fairly profound effect on any human who is not safe behind the nearest fence, and a deer doing the same thing is scarcely less dangerous. It is a long way from the world of Bambi, and a great deal nearer the truth. In the dusk, even a wild stag may mistake you for an adversary and *any tame or semi-tame deer should be approached with extreme caution in the rut.*

A curious form of aggressive display is known as *parallel walking,* where two stags will parade up and down alongside one another, separated only by a few yards. The whole thing looks extremely formal but no doubt signals are exchanged which we find difficult to understand. The parade may break off inconclusively or turn instantly into a savage fight in which the two stags lock antlers like a rugby scrum, pushing with all their might to make the other yield ground. If one of them is forced backwards or trips, he will be lucky to escape by ignominious flight, quite likely aided with a vicious stab to the flank or the backside as he turns. Because the antlers are locked, the contestants are more or less nose down, and during the fight ground is likely to be given and taken by both sides. One may see evidence of this by long scores in the turf if one or other of the fighters has long brow points. The clash of antlers as they come together can be heard a long distance off.

Even the little roe is a fierce and determined fighter, and I have watched two bucks hard at it for nearly a quarter of an hour before one was prepared to acknowledge defeat and make off. With all the pent-up energy of the rut, the speed and savagery of attack and the clever way they use their antlers to fend off the other is marvellous and somehow terrifying to watch. While most fights end in a moral rather than a physical victory, serious injuries do occur, mostly at the moment of disengagement or if one deer slips or trips and finds himself at the mercy of the other. Muntjac tend to use their tusks rather than their antlers and most mature bucks carry scars on their necks and ears from past fights.

When the rut is over there will be a number of walking wounded; animals that limp, or have broken antlers. Others, which one can tell from their hunched appearance or their lack-lustre behaviour, have received some internal injury which may in the end be fatal. Deaths also occur from time to time as a direct result of fighting, and on occasions the contestants may become locked together by the antlers and unable to free themselves. This miserable state continues until they both starve or by frantic efforts one either frees himself by snapping an antler or breaks the neck of the other, in which case he will he anchored by his opponent's corpse until he too collapses. It is easy to imagine how animals with complex antlers, such as the red or fallow deer, could become locked together, but even roe deer can do the same and I have known of nine cases where death was the result. Usually they are large and well-matched bucks with well-developed and well-pearled antlers, but on an estate near Salisbury a very large buck was found enmeshed with a very much younger four-pointer buck which one would not have expected even to measure up to him. Another pair became fixed on either side of a tree and so they were completely unable to move and were only found after suffering what must have been a prolonged and very painful death.

The peak of the rut

The idea of the male rounding up his females like a sheep dog comes from observations of red deer on the open hill in Scotland. This is not typical behaviour, even of red deer. It is more or less unique to the Highlands. People who are used to forest-living red deer on the Continent are astonished at the bellowing of half a dozen stags echoing all day among the hills and, equally, at the sight of a master stag with fifty or sixty hinds which he has to defend night and day by constant vigilance and threat; running this way and that, never having time to feed. No wonder that after a fortnight or so of this hectic activity he is exhausted or *run*. The stag has used up all his reserves painstakingly built up during the summer and must retire from the fray to recuperate from a kingsized hangover, while previously unsuccessful rivals take over. If he is lucky and undisturbed by thoughtless people he will be able once again to feed and regain weight before facing the rigours of winter. Not all manage to do so and their carcasses feed the hooded crows, ravens and golden eagles in nature's ruthless logic.

The rut among forest red deer pursues a very different course. Little roaring will be heard except on frosty nights, and the old stags will

scarcely utter more than an occasional deep grunt. There is no need for the hectic activity and challenges of deer on the open hill. Most red deer in the forest will be found in small family parties consisting of a stag and three or four hinds with their followers. Because of the trees, one stag is unable to see his neighbour and does not feel the need to start roaring in challenge. The frosty-night choristers are mostly young hopefuls who have not yet achieved domestic bliss.

With fallow the stand buck's groan, a noise which has been described as a rhythmical belching grunt, is more difficult to explain because, unlike the red stag, a fallow buck grunts because he is on his rutting stand and not because another buck is challenging him. Except where the rutting stand is in a valley or other favourable acoustic situation, the noise does not carry very far. Round the outside of the charmed circle will probably lurk a handful of young bucks, while a dozen or more does attracted to the buck will drift about and feed when not actively involved in the rut.

Listening to the sika rut is an eerie business of dusk and dawn and moonlit nights in the forest. Sika are more nocturnal than either red or fallow deer, and their rutting behaviour seems less tied to a particular spot. A sika stag will certainly choose his area of woodland, giving vent to his loud triple whistle every ten or fifteen minutes, presumably to advertise his presence to any hinds which may be in the area. Sika too are quick and determined fighters and the same ring of young hopefuls is likely to be encountered round the master.

The rut among territorial deer

Presumably because they lead much more solitary lives, the three smaller species, roe, muntjac and Chinese water deer, are much less obtrusive in their rutting behaviour. In fact, it has been said of the latter that one hardly knows that the rut is on at all, there is merely an increase in 'wickering and squeaking'. The muntjac, of course, is a non-seasonal breeder and so rutting, which follows immediately after the birth of the fawn, is an individual performance and the buck will be in close atten-dance on his doe for a week or two before parturition. Fights among them, as in the case of roe, are more often to do with territory than with competition for the favours of a doe.

At the onset of the roe rut the bucks do show an increased tendency to bark at one another, while any unaccompanied doe will look for a

A young red stag 'fleering' – lifting his lip to expose the scent organs of the upper gum.
Photo Brian Phipps

buck and bring him back to the place where she wants to rut. This she does by flirting with him, which can be very amusing to watch. She makes a shrill squeaking noise to attract the buck and induces him to chase her at increasing speeds as the tempo of the rut builds up. Sometimes these chases take the form of a tight circle round a bush or stone or similar object, until a small path is formed called a *ring*.

The buck makes a rasping noise in his throat and it would be easy to assume that he was chasing the doe, but in fact the reverse is the case, the doe is encouraging him to chase her! This is clear when the buck, as often happens, flops down exhausted. The doe will immediately stop running away and will sometimes attempt to get him up again by striking at him with her forefoot.

It is probable that much of the rushing about which is a feature of rutting behaviour in many deer species is a way of reconciling the female to being touched by another animal. Unlike cattle, deer are not contact animals, and once past adolescence the only urge to touch one another is between mother and young, apart from the period of the rut.

You will be lucky to see the actual act of mating, although the female once she is willing may be mounted many times before a successful mating is achieved.

Consideration for rutting deer

The rut is such an exciting time to watch deer that sometimes people forget that only the drive to reproduce makes them relax their normal caution. It is grossly unfair to abuse this by needless disturbance. By all means watch the rut, but watch it from a distance; use binoculars or a telescope and do not try to get too close, or the alarm will be given and rutting activity will have to be deferred. Do not think that shooting is the only way of disturbing deer. The sudden arrival of a human, armed or not, in the middle of a rutting stand is immensely alarming and if repeated too often will at least make the deer shy and unapproachable.

The greatest enjoyment to be had from deer watching is to observe deer about their normal life, and this means that they must be unaware of anyone nearby.

5 • *Watching Fallow and Other Park Deer*

If you are only prepared to sit down and be patient for an hour or two, there is no better place to study deer than in a park. Most people try to get too close to the deer and disturb them. Photographers are far the worst offenders. Bring a pair of binoculars, find a vantage point under a tree, and just wait until the deer forget all about you and resume their daily life. Just a few visits to a park through the seasons will give you an insight into their way of life that would have taken years of patient study among wild deer. Their most intimate secrets are revealed for all to see in the confines of park life: rut and fawning, fighting and suckling, the care of a mother for her young, the sadness of an old buck displaced by a younger and stronger adversary. The sight of a score of fawns playing 'King of the Castle' round a massive tree stump is unforgettable, and even when the deer are resting there may be a magpie catching flies round their heads, or a pair of jackdaws cheekily pulling beakfuls of fur from their backs and flying in relays to their nesting hole in a neighbouring giant oak.

The landscaped beauty of a park and its general air of changeless repose are refreshing after the bustle and noise of the city. This is no new idea, for five hundred years before the time of Christ, Buddha chose to give his first discourses in the tranquil atmosphere of a deer park.

The first step, obviously, is to scan the deer and decide what species are present. Now that there is so much more interest in deer parks there is quite likely to be more than one species and maybe some oddities like Soay sheep, or even the old breed of English park cattle which are white with black noses and ears.

Although fallow are the commonest species to be kept in a deer park many contain the larger, unspotted red deer. Sika, either the Japanese or the larger Manchurian race, also exist in several parks. Sika can be mistaken for fallow because they are heavily spotted in summer coat, but if you see a sika stag it will have rounded antlers in contrast to the flattened palmated antlers of a fallow, and they have a V-shaped white mark on their foreheads and a prominent white spot on the outside of the hind leg just below the hock.

Watching a red stag in the rut. This is a potentially dangerous animal and should not be approached closely. Note the group all have stout sticks! Photo Brian Phipps

Occasionally you may visit a park which has some of the large odd-looking and ungainly Père David's deer; specialist collections, like those at Woburn Park, will have other varieties to test your deer knowledge.

Fallow deer have been gracing British parks since the days of the Normans. They are ideal park deer, intermediate in size, statuesque and beautiful, the final ornament to the approaches of a great house. To the deer watcher they may be confusing because, as can be seen from the fact sheet they come in different colours. One may come across all shades from pure white through cream, sandy, brown, silver, pied and black, with as wide a variety of spotting. One has almost to disregard the coat and look at the animal itself, its shape, size, expression, tail, and shape of antler, to make sure that it really is a fallow.

Some of these colour varieties are supposed to be regional in origin. For example, the black fallow from Epping Forest were said to originate from some of this breed which King James VI and I brought from Norway to Scotland, and afterwards transferred to Epping Forest. Even if this story is true there were other fallow already present in Epping at the time. The New Forest had the same sort of tradition. The fallow there at one time were all common-coloured, a few being heavily spotted and with occasional white animals in addition. In fact, the maintenance of the herd with one colour predominating depends very largely on the keepers. There was by tradition a prejudice among them in favour of dark fallow at Epping, and equally in favour of common-coloured fallow in the New Forest. As soon as this selective tradition is allowed to lapse a mixture of colour develops. In fact, many of the Epping fallow are not black now, and in the New Forest a variety of colours has developed, as has always applied to the deer in the surrounding woodlands. Many of the old deer keepers used to say that the colour of a fallow fawn was determined by the sire not the dam. Geneticists would probably dispute this but there was an occasion years ago when a black buck jumped into a Dorset park and joined a herd of fallow that had, up to that point, been uniformly menil, the variety which is a bright chestnut in ground colour with bright spots throughout the year. The black buck's days were numbered but he obviously used them to good effect because to this day one or two black fawns are born each year although to preserve the traditional colour none of them is ever allowed to reach breeding age.

In the Victorian heyday of the great houses, owners vied with one another to show the greatest variety of animals in their parks. Some, like

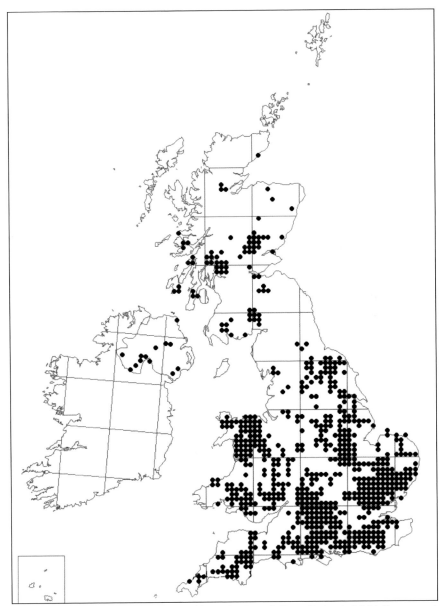

Fallow Deer Distribution in the UK from the British Deer Society's 2001 Survey and reproduced by kind permission.

the Duke of Bedford at Woburn, accumulated an astonishing collection of exotic breeds of animals and birds which included buffalo and the amazing brush turkey that makes a compost heap and leaves its eggs to incubate in the heat created. It was the Duke, incidentally, who was

FALLOW DEER (*Dama dama*)

Male: buck. Female: doe. Young: fawn.

LENGTH:	170cm (67in).
HEIGHT AT SHOULDER:	Buck: 90–95cm (35–37in). Doe: 80–85cm (32–34in).
WEIGHT:	Male: *c* 70kg (160lb) upwards depending on condition. Female: *c* 45kg (95lb) live weight.
PELAGE:	Very variable. Types usually found: Common: chestnut with white spots in summer; winter, dark brown and unspotted. All year, light-coloured area round tail, edged with black. Light tail with black stripe. Black: all year, black shading to greyish-brown. No light-coloured tail patch or spots. Menil: spots more distinct than common in summer with no black round rump patch or on tail. In winter, spots still clear, on darker brown ground. White (not albino): fawns cream-coloured, adults become pure white, especially in winter. Dark eyes and nose. No spots.
ANTLERS:	Mature bucks have typically palmated antlers. Shed April–May. In velvet to August.

BREEDING: Rut late September to late October. Fawns born June. Twins rare.

DISTRIBUTION: W Europe up to 60° latitude; Asia Minor. Elsewhere by import and escape. England: widely distributed especially Midlands and South Wales. Less plentiful and local. Scotland: mainly lowlands, Perthshire, Argyllshire; also Islay, Mull, Scarba. Ireland: widely distributed.

HABITAT: Ideally, deciduous or mixed woodland interspersed with farm land. Plantations in thicket usually avoided.

FOOD PREFERENCES: Mainly grass, cereals, herbs, fruit, nuts and berries. Some browsing, especially hardwood shoots. Bark taken from smooth-stemmed hardwoods.

HABITS: Separate herds except in autumn. Vagrant over home range, depending on food and disturbance. Moves fast with 'pronking' gait (all four feet moving together). Most feeding dusk and dawn on fields. If unmolested may feed at intervals all day. Old bucks very secretive and nocturnal. Juvenile bucks have play rings.

VOICE: Does and fawns communicative, with high piping nasal cry. Does have gruff alarm bark. In rut mature bucks 'groan' – a rhythmic belch.

IDENTIFICATION: Distinguished from red deer by smaller size, long tail, often spotted. Bucks have flattened antlers. Sika have white V-mark on forehead, white glands below hock, rounded antlers and predominantly white tail. Roe and muntjac are much smaller, no spots. Roe has no visible tail, muntjac has pig-like silhouette.

responsible not only for saving the Père David's deer from extinction, but for providing the springboard at Woburn for the escape and eventual establishment in the wild of our muntjac and Chinese water deer.

Other owners prided themselves on the size and variety of their fallow, and any new colour variation was eagerly bred up. That notable eccentric, Walter Winans, listed no fewer than ten varieties of which the rarest were black with white legs, and black with white spots. Sadly these seem to have died out completely but one may occasionally come across fallow of a bluish colour and others with parti-coloured patches.

A very interesting coat variety of fallow is to be found in Mortimer Forest in Shropshire. Apparently this arose as the result of a mutation among wild fallow as it was not noticed until the 1960s when some individuals were seen to have a long shaggy coat, the winter coat being made up of hairs about three times as long as is usual. There are long curly hairs on the forehead, and buff or ginger hairs up to 13cm (5in) long flow from the ears. Long hairs on the tail almost double its apparent length. Possibly because of selective protection this type grew to represent about 25 per cent of the fallow in the area.

In spite of their statuesque look in the park, fallow are one of the more nervous of the deer species. When alarmed they form a bunch, a

Fallow come in several colours. The white buck is not albino. His companion is common-coloured, chestnut in summer but darker in winter with a whiter belly.

defence mechanism which was probably developed when they were preyed on by wolves. When the herd breaks and runs, a lead doe apparently makes the decision. She will be followed by the bulk of the does and fawns with any bucks in the group at the back. Some park herds are much wilder than others; some will throw their heads up at almost every noise or movement within the park, while others are more placid. The fallow in Knole Park near Sevenoaks in Kent are remarkable for their strong nerves but maybe the presence of a golf course in the middle of the park has something to do with it.

Ideally, part of the park should be bracken-covered or have thickets which the deer like to use as cover when they are disturbed or when the flies are bad, and where they can hide their fawns. There should be a rule in every park that all dogs should be kept on leads at all times and totally excluded during the old Fence Month, 9 June to 9 July, when the fawns are being born.

This natural timidity of fallow contributes to their suitability as a park animal because unless hand-reared they are unlikely to be really tame. Some people might think this a pity but, in fact, any deer which has lost its fear of man may be extremely dangerous, especially in the rut. Even females can inflict very nasty wounds with their feet which they use like flails and, of course, the stags and bucks are potentially lethal unless they retain a degree of respect for humans.

I have seen people taking the most horrid risks where public rights of way run through a park. Getting between a red stag in full rut and his hinds, for example, is foolhardy in the extreme, and it is inviting trouble to offer them food at any time. I was attempting once to take a photograph of a rutting stag in Richmond Park and in case of trouble had armed myself with a stout stick. Having made a careful approach behind an ancient oak tree while the animal was roaring and throwing up bracken with its antlers, I peered round the tree to see to my horror that a nursemaid with a toddler and a pram was encouraging the child to offer it a biscuit. Not long after this one of the BBC television announcers received a face wound which needed thirteen stitches for doing the same sort of thing. You need to be as circumspect with a

The spotted fallow are menil. Note the darker coat colour of the sika hind with them.

tame deer or with park deer in the rut as you would be with a dairy bull.

Many people have attempted to rear roe fawns in the mistaken idea that they have been abandoned. The few that succeed find at the end of the year, should the fawn be male, that they have nursed up an animal which is difficult to feed, destructive to its environment, and apparently filled with hate of its captors. Because it is so small and so nimble there are few animals more dangerous than a 'tame' roebuck.

If fallow are timid they can also be inquisitive. There was a summer girl guide camp in a West Country park and one night the fallow bucks started playing with the tent guy-ropes. Of course, some pegs came out of the ground and very soon there was consternation and turmoil. This naturally frightened the fallow and one buck became entangled in a tent and bore it off on his antlers. A search was made but the tent was not found until a few days later when the unfortunate buck was discovered still wrapped in canvas but drowned in one of the ponds where he had fled in his panic.

Even disregarding the other reasons for not leaving rubbish anywhere, one should be particularly careful not to leave items around in the deer park which the fallow can either get wound round their antlers or eat: plastic string, coils of wire, wire netting and polythene sacks have all been the cause of alarm or death. Tins may be trodden on, injuring the feet, or they become attached to a deer's nose. Some extraordinary items have been recovered from fallow stomachs, including several complete items of underwear, so make sure you dress before you leave the park.

Antler growth

If you are lucky enough to be able to visit a park regularly it is fascinating to single out one individual buck, who may be recognisable because of his unusual colour, and study the development of his antlers from the time the old set falls off in the last half of April to the time that he proudly burnishes his new set ready for the rut sixteen or seventeen weeks later.

The ugly scar left by the shed antler soon heals over and two mushroom-shaped humps of velvet, grey or pinkish depending on the animal's coat colour, soon appear. By early June, elongation has started and one can already see the front stubs which will grow into brow points separating from the main beam. By the end of that month the second point should have appeared and a club shape will be forming at the end of the antler which is the start of the palm developing. By early August growth should be nearly complete and the velvet, under hormonal influence, starts to decay, often attracting swarms of flies which cluster to form dark masses on the antler. The deer are always twitching their ears and their heads because of the irritation.

The young bucks will be first to shed their velvet, followed a week or two later by their seniors. They will search the park for low or fallen branches on which they rub their antlers until the velvet peels away leaving white bone underneath. The process of cleaning takes a few days and while strips of velvet still hang from the new antlers, the bucks are known as being *in tatters*, a very descriptive term. The deer themselves can be seen to chew discarded pieces of velvet and the jackdaws and magpies may also filch a strip for their own purposes.

Burnishing the antlers continues until they are polished and brown. A well developed mature fallow buck should have antlers with broad palms with a series of small points, the *spellers*, on the posterior edge, well developed brows and *trez* (tray) points. Some bucks grow palms which are divided so that they have a fishtail appearance. This is regarded as a bad fault. A fallow buck reaches the peak of his development between seven and nine years of age when in the old language of venery he would have been called a great buck. In the years preceding he would, in turn, have been a fawn, a pricket, a sorrel, a sore, a barebuck and a buck: a long apprenticeship for two or three short years of supremacy.

While you have been concentrating on the antler growth of your chosen buck, the daily life of the park will have ebbed and flowed round your vantage point. June will have seen the birth of the fawns, and nothing is more touching than to see their first staggering steps. Soon the evening air will be resonant with the nasal bleat which mother and young use to keep contact with one another when the herd is on the move.

Fences, leaps and outliers

Should you be allowed the full freedom of the park – and not all park owners allow the public to roam at will – there are many interesting things to be learnt by walking round the fence. You may have noticed as you entered the park that there was a grid to allow traffic to pass but not deer. Fallow in particular are very clever at getting over these grids and in more than one park I have seen their dainty slot marks where they have trodden between the bars onto a narrow girder below.

You may be lucky enough to find remnants of the attractive old paling fences which used to surround many parks. They were made of oak which was cleft, that is to say split along the grain, not sawn, which was stronger and longer lasting. Each district or park had its traditional pattern of paling, usually a combination of long and short staves making smaller gaps at the bottom to prevent the fawns from getting out, and a slightly irregular top which was supposed to be less easy for deer to jump than a fence with a continuous top. Deer leaps were incorporated into the fence so that any *outliers*, deer that had escaped or bred outside, could jump back into the park. Look for a place where the ground is higher outside and where the fence leads into a low wall. The jump down is not too daunting but because the ground falls away, leaping up the wall from the park side and balancing on the bank at the top is beyond the strength of an animal trying to get out.

Fallow are, of course, very widely distributed, particularly in England, and practically every park will have its outlying deer, either because they have escaped or because they are attracted by their cousins

inside the fence. With permission from the landowner, looking for these outliers round about a park is an excellent introduction to the challenge of studying fallow deer in the wild. You will need all the deer craft that you have learnt by studying what goes on in the park to get on terms with these very wild animals. They are not only well able to look after themselves but have been doing so in many cases since the Middle Ages, with the hand of man invariably turned against them. An old wild fallow buck is one of the most crafty animals you could wish to try and outwit. You can find his bed where he lies up, trace his paths where he goes out to feed; find the bushes he thrashed with his antlers, and even hear his deep belching groan in the rut, and yet when it comes to recording what you have actually seen on all but very exceptional occasions, it will be does and fawns with the occasional young hopeful. Methuselah will have vanished like a puff of his own rank odour.

You have to learn the technique of looking continuously all around and yet never treading on a stick, the trick of passing so slowly through the woods, and with such concentration, that the fall of a dead holly leaf makes you jump, and always to be conscious of the shifts and eddies of

The magpie sitting on a hind's back is probably searching for ticks, or stealing hair for its nest!

Watching park deer reveals all their family life, much more difficult in the wild.

wind, knowing that one false move will bring your scent to the deer and stalking that group will be over for the morning.

Wild boar

The wandering deer watcher just may encounter a wild boar these days and so it may be appropriate to set out a little about them and how to react if one comes your way.

Wild or hybrid pigs have been farmed in Britain for some years but it was not until the great gales of the late 1980s and 90s that the fences designed to keep them in proved inadequate. Some escaped and soon feral populations became established. Boar are great truants from captivity – they jump like deer and burrow like moles – and the conditions in Britain suit them fine. They were, after all, resident here until Tudor times. The main centres of occupation are on the Kent–Sussex border, and in Dorset, with other reports regularly coming in from many areas. It is very likely that boar have joined our resident fauna for the foreseeable future and will increase their distribution. There are

In some parks you may come across the strange, ungainly Père David's deer which came from China. Photo Brian Phipps

worrying features, especially for farmers of domestic pigs, but any threat to the general public is minimal. However, a sow with piglets could turn on an intruder and a big male especially if hurt in any way can be a dangerous beast with his long, razor-sharp tusks. The rule should be to leave them well alone.

Like all pigs they are great diggers and the most noticeable sign of their presence is evidence of heavy rooting in the fields. Badgers do this too, but on a lesser scale. Casual encounters are unlikely except for sighting a boar in the car headlights as they are extremely shy and wary. With good local information, the best chance of seeing them is to sit in a high seat overlooking a field where they have been recently rooting, but be prepared for a long wait and probably a night vigil because they are more nocturnal than deer.

Photos of wild boar make them look ugly with their rough hair and generally muddy appearance, but in their natural habitat the picture is completely different. In addition they are brave and intelligent, so they make an extremely interesting addition to the countryside.

How NOT to do it! Even when in early velvet, like this sika stag, over-tame deer can cause injury. Deer are not pets.

Sleigh ride

The reindeer was once abundant in Britain. Its bones have been dug up from Trafalgar Square in London and from the very foundations of the Natural History Museum in Kensington. After the last glaciation reindeer returned and it was mentioned in the *Orkneyinga Saga* that the Earls of Orkney used to hunt them regularly in Caithness in the twelfth century. They must have disappeared soon after.

In the 1950s Mikel Utsi brought some of his Swedish mountain reindeer to Scotland and established them in the Cairngorms. More followed, some of forest type and some from southern Norway. These deer bred and their descendants can be visited high on the Cairngorms in the Glen More Forest Park near Aviemore. It is one of the most unusual and exciting deer watching expeditions that one can imagine. Accompanied visits to the Cairngorm herd are usually arranged daily. Normally a guide leaves Reindeer House above the Glen More camp site for the hill at 11.00am. They have a good web-site (http//www.reindeer-company.demon.co.uk) which gives plenty of detail.

Deer parks open to the public

Deer watching opportunities in parks, zoos, deer farms and elsewhere, can be found throughout the country where not only deer are to be seen, but where deer watchers are actively welcomed. Search Google (google.co.uk) for 'deer park' or check with the National Trust (nationaltrust.org.uk) to find those nearest to you. The National Trust Handbook is a mine of useful information. Do not rely on free access to a park at all times. A check should always be made before a visit.

Other parks exist where you may be welcome with previous permission. In any case it is best to check first. By doing so you may easily get VIP treatment!

6 • *The Rise and Fall of the Deer Parks*

Many of the deer parks scattered up and down the country are real relics of the Merrie England of Robin Hood and Good Queen Bess. Ever since the arrival of the Normans, attitudes to deer in England have been affected by their passion for the chase. The old idea, dating back to Roman law, was that game belonged to the man who killed it irrespective of land ownership. This notion, already in decay among the Anglo-Saxons, was swept away by the Norman conquest but lingered on in Wales and Scotland where to this day a poacher is quite likely to stand up in court and talk about the honest man's right to a stag or a salmon.

As Norman influence spread, the forest laws were quickly imposed which prevented commoners from hunting, and reserved vast areas of the country as royal forests or hunting preserves in which the most severe penalties, including death and blinding, could be imposed for poaching. The Beasts of the Forest were four: red deer, fallow deer, roe and wild boar, although the hare was often included and was held in very high estimation as a quarry species. Roe deer were reduced in standing to Beasts of Warren in the fourteenth century. Although most of England at that time was very well wooded, the word 'forest' was used more to designate the royal hunting preserve than purely woodland. One may still, for example, see Exmoor Forest marked on the map although it was never particularly tree-covered, and similarly the New Forest comprised large areas of heath in addition to the woods. Traces of former royal forests often can be found in place names (Buckholt, Kingstag and so on).

If the king granted hunting rights to one of his nobles that area was thereafter known as a *Chase* (Whaddon Chase, Cranborne Chase) and the severity of the Forest Law was replaced by milder penalties, although justice even then was savage and arbitrary according to modern ideas, often involving confiscation of the offender's entire property, livestock and belongings:

> The said foresters on the authority of Geoffrey of Leuknore, came to Martin, to the house of John the Bor, and there by force took a certain man of Forton and, accusing him of a felony, led him to Cranborne in the county of Dorset and there at their own will and without reason hanged him. And having hanged him they came to Bredeme [Bridmore] in the county of Wiltshire and there took two cows, from the cows of Peter of Skidemore, saying that they belonged to the felon that was hanged. By force they took those cattle to Cranborne and detained them there. And they went to the house of Richard of Martin and took a horse belonging to the man who was hanged, value at ten shillings, and they found in the

home of the hanged man a silver buckle, valued at 8d. (Taken from *The Hundred Rolls for 1275*)

The felony concerned was probably poaching although it may equally have been to do with *vert* rather than *venison*, that is to say, cutting down a tree without authority.

The idea of containing animals within a park is of great antiquity. Parks existed throughout the Roman Empire, including Britain, while the Domesday survey (1086) lists thirty-five parks existing at that time. Soon there were to be many more. The idea of these early medieval parks was almost certainly to provide an easy source of meat for the winter. Until the cultivation of root crops was developed in the eighteenth century, providing enough food for domestic stock to last them through the winter was extremely difficult. All surplus stock was killed and salted down in the autumn. Fresh meat from wild game must have been highly appreciated.

Because the right to hunt and kill game was a royal prerogative, the creation of parks into which wild deer were encouraged to jump was only permitted by royal licence, or in the case of a park within a Chase, by permission of the owner of that right. The park was usually created by digging a ditch and erecting a fence of wooden palings on the bank

The traditional scene: a herd of dark fallow deer in front of Parham House in Sussex.

A forest of antlers. Fallow deer are the time-honoured species to ornament a gentleman's park. Photo Brian Phipps

of earth excavated. Deer were encouraged to jump in by the provision of 'deer leaps', gaps in the fence where deer can jump down, but where the ground is too steep to allow them to jump out again.

By careful use of a map and by keeping your eyes open when walking in the country, relics of these old deer parks can often be traced. For example, there is one near my home called Harbins Park, in Cranborne Chase, which was the scene of a violent argument between the owner of the park, Mr Harbin, and the proprietor of the right of chase, Lord Rivers:

> Mr Harbin inherited the estate from his ancestor and peaceably enjoyed the privilege of the enclosure as his ancestor had done, and might have continued to do, had he not made an unfair use of it by converting some of the pales on the Chase side into a sort of pitfall, so that the deer could easily leap in but not get back again, and to induce them to be thus entrapped they were enticed by apple pomace of which the deer are particularly fond. (Chafin *Anecdotes of Cranborne Chase* 1818)

Out hunting, Lord Rivers discovered this abuse of his privileges and caused his servants to throw down the pales of the fence. The park, it appears, was never remade after this, but the earth embankment can still be traced by those with an eye for country.

Another type of enclosure which can sometimes be discovered was the 'haye', a deer trap into which they could be driven and more easily killed. Most hayes were in Cheshire, Shropshire and Herefordshire where, once again, local names, possibly of woods or fields, can betray their original location. In the south, thick hedges (haye means hedge) or paling enclosures were used to restrain the deer. In the barren uplands, similar traps were made using stone walls, and traces of some of these can still be found. There is one not far from Shap in Cumbria, in a valley called Wet Sleddale. Another exists on the Island of Rhum. These northern deer traps were referred to as 'settis' (seats) and like the hayes were used as a convenient way of filling the larder on less ceremonial occasions than the large-scale communal hunts called tainchel.

An item in the old *Statistical Account of 1796*, described the former use of one of these traps:

> Before the use of firearms their method of killing deer was as follows: on each side of the glen formed by two mountains, stone dykes were begun pretty high in the mountains and carried to the lower part of the valley,

A deer leap allows deer which have escaped to jump back into the park, but not to jump out.

always drawing nearer till within three or four feet of each other. From this narrow pass a circular space was enclosed by a stone wall of a height sufficient to confine the deer. To this place they were pursued and destroyed. The vestige of one of these enclosures is still to be seen in Rhum.

A little research and knowledge of the history of our deer can so often give additional interest and point to a ramble in any country.

The deer that the Normans enticed into their parks would at first have been the native red and roe deer, although the latter never take kindly to park conditions. Soon after the Conquest, however, fallow deer start to be mentioned, and indeed were listed as one of the Beasts of the Forest. Having died out during the last glaciation, fallow were re-introduced into Britain by human agency, some time between the Roman era and the Middle Ages. It is known that they had been brought to large parks in Gaul by the Romans in the first century AD and nothing would have prevented their import to Britain during the long settled period of Roman rule in this country. However, no solid evidence of their presence has come to light, in spite of the enormous

Some cover is needed in any deer park to shelter new-born fawns and to give shade and some concealment.

The deer leap at Wolseley Park, showing cleft oak paling which wasd the traditional fencing used for deer parks. (From Shirley's English Deer Parks)

number of excavations of Roman sites. Until the vital authenticated bone or antler emerges, one has to assume that it was the Normans who were responsible. Domesday Book has nothing to say on the subject, but ancient records after this date begin to speak definitely of them. Although at first they would have been strictly confined, fallow are great escapologists and no doubt it was not very long before they were breeding in the wild. By the twelfth century the forest which lay to the north of London, according to Fitzsteven, was 'well stocked with red stags and fallow deer'. By 1223 they were sufficiently numerous in the New Forest for William Briwer to be allowed to chase bucks there with his hounds.

As well as providing meat, the medieval parks, some of which were very large in extent, were used for hunting. Woodstock Park, Oxfordshire, was 400 hectares (1,000 acres), while Hulme Park, Northumberland, originally contained 1,400 hectares (3,500 acres).

During Tudor times the deer park took on a new dimension as an ornament and setting for the great houses which were then being built. The spreading lawns and thickets of a great park enhanced a display of opulence and offset the geometrical precision of parterres and formal gardens. The deer themselves, picturesque and a symbol of sport and leisure rather than of industry, completed the landscape. Hunting continued but by Elizabethan times, on occasions at least, degenerated to ceremonial. Queen Elizabeth I liked to sit in a bower, past which the choicest stags and bucks were driven. These she shot with a cross-bow. Apparently she was very accurate. Her hunting lodge in Epping Forest can still be seen.

By this time the nature and appearance of a deer park as a place of beauty and contained wildness was already well established. Shirley in his account of English deer parks quotes from a book written in 1616:

> The parke would be seated, if it be possible, within a wood of high and tall timber trees, in a place compassed about and well-fenced with walls made of rough stone and lime, or else of bricks and earth lome or else with pales made of oak planks. Nor ought the parke to consist of one kind of ground only, as all of wood or grass or all coppice, but of divers as part high wood part grasse or champion and part coppice or underwood or thicke spring. Neither must the parke be situated upon any one entire hill, plaine, or else valley, but it must consist of divers hills, divers plaines and divers valleys. The hills which are commonly called the viewes or discoveries of parkes would all bee all goodly high woods of tall timber as well for the beauty and gracefulnesse of the parke but also for the echoe and sound which will rebound from the same when in the times of hunting either the cries of the hounds, the winding of hornes, or the gibbetting of the huntsmen passeth through the same, doubling the musicke and making it tenne times more delightful.

These days, when tranquillising drugs and rapid transport make the capture and resettling of livestock much easier, one may overlook the difficulties which park owners in medieval days had in stocking their parks, once the local deer had disappeared, or in introducing new blood. The royal parks in particular were not self-sufficient because of the demands of the royal kitchens and annual gifts known as the Royal Venison Warrant. In fact, large numbers of deer have for centuries been shipped about the country and one might wonder how it was done.

Catching was usually by means of strong nets known as deer toils, into which deer were driven and entangled. Anyone who has been involved in catching deer in nets for research will know that this can be not only a very skilled, but a very hairy performance in more ways than one. The official in charge of this for the royal parks was known as the Yeoman of the Toils.

Trained dogs were also used to catch park deer and hold them down

Fallow deer led by the power of music.(From Shirley's English Deer Parks*)*

without injury until they could be restrained. Once caught, how were they to be transported along the unfenced quagmires that passed for roads in those days? Tradition has it that the roe that were reintroduced to Dorset around 1800 were brought from Perth in 'padded horse-boxes', but the majority of park deer must have been driven, presumably with the aid of cattle dogs. A curious tradition persists that certain individuals practised the lost art of moving deer with the aid of music, like the Pied Piper of Hamelin.

> Travelling some years since I met on the road near Royston a herd of about twenty bucks following a bagpipe and violin which, while the music played, went forward. When it ceased they all stood still, and in this manner they were brought out of Yorkshire to Hampton Court. (Playford, *Introduction to Music* quoted in Shirley's *English Deer Parks* 1867)

From a peak of nearly 2,000 in the mid-thirteenth century, the number of parks declined steadily. In Elizabethan times there were perhaps 700. Many of these were ravaged during the Civil War and some were never re-established. By 1892 395 remained stocked with deer in England, and by 1949 this number had fallen to 143, with a further thirty-four in Scotland, Ireland and Wales. Twenty years later only 101 remained

A red stag and his hinds in the tranquil surroundings of a deer park. Photo Brian Phipps

active, although a welcome revival in interest was already apparent by that time. In 1985 there were once again 240 active parks, although this figure includes some wildlife and safari parks and a number of deer farms where they are kept under more intensive conditions. Of the traditional parks, forty have been lost since 1949 but at least eleven of those previously abandoned have been revived.

It is a tribute to the sense of responsibility of succeeding generations of park owners that so many have been preserved and kept stocked with deer in the face of the ruin of two world wars, punitive taxation, and the enormous cost of maintaining walls and fences which are often measured not in yards but in miles. The efforts of the National Trust and others to preserve our heritage of great houses fortunately extends also to the surrounding parks in recognition of the contribution which a deer park alone can make to the landscape. In Richard Jefferies' words: 'A park without deer is like a wall without pictures'.

7 · *Muntjac – Invader of the Suburbs*

The Barking Deer, or Reeves muntjac, may already be England's most numerous deer species. Yet how much do we know about them? Sadly, the clearest evidence of the presence of this small deer is the all-too-common sight of a huddled form on the side of the road where a muntjac has crossed at the wrong moment – and has paid the price. When disturbed in the woods they can be easily mistaken for a fox or a small dog. When they are really running, muntjac are long, low and brown, with a noticeable tail which is sometimes carried vertically. When feeding in the fields their hunchbacked silhouette makes them look rather pig-like. In fact, they are highly individualistic in appearance but this is only obvious to the casual observer at close quarters.

The buck carries short, simple antlers on long furry pedicles which are extended as ridges of bone between the eyes. These ridges are further emphasised by a line of dark hair. At one time muntjac were called 'rib-faced deer'. In the depression between these two ridges is a pair of large glands *(frontals)* enclosed in a fold of skin; another pair of glands lies below and in front of the eye *(suborbitals)*. Both pairs gape

83

open at times of emotion or stress. The sockets in the skull which house the suborbital glands are as large as the eye sockets and not only help to identify a muntjac's skull if you happen to find one, but indicate too the importance of scent in their daily life.

Another unusual feature is the development of the canine teeth, especially in the male, to form tusks which in mature bucks jut down below the lip. They are used for fighting among themselves – many bucks carrying typical scars as a result – and also to fray small plants in place of using the antlers, and as a weapon of defence. The larger Indian muntjac has a reputation for injuring dogs when cornered, which was supposed to have been the reason why the original colony of this species at Woburn in Bedfordshire was eliminated in favour of the Reeves. Even they are capable of standing up for themselves. Phil Drabble, that acute observer of wildlife, wrote of one of his tame muntjac:

> The badgers which love young rabbits come to a feeder by the house in the evening, and we were worried that they would prey on the young muntjac which was in rushes one hundred yards away. Our fears were groundless. A floodlight enabled us to watch the whole drama. The muntjac doe arrived when a husky boar badger was feeding. Although only half his weight she saw him off and followed up her attack for a hundred yards or more. He did not stay to argue.

Muntjac put their tusks to such vigorous use that they often get broken. Instances have been recorded of a doe attacking a dog on a lead, using forefeet and teeth and of a captive doe attacking a human, who was returning an accidentally separated fawn, by flicking out with her forefeet and biting. I went to see a friend who had some tame muntjac in his garden. The wellies which he put on before going into their pen were slashed in several places, which he claimed was done inadvertently but giving an indication of how the teeth are used in fighting. The very possession of tusks in association with simple antlers proclaims the ancient lineage of the muntjac. Their fossils have been found in Miocene deposits in France and Germany dating back between fifteen and thirty-five million years, long before more modern deer evolved.

If you do find a muntjac skeleton, all too common unfortunately because of their liability to road accidents, examine the bones of the legs and feet where more primitive features will be apparent: in addition to the two halves of the foot, or cleaves, there are two dewclaws which are remnants of the second and fifth toes. Above them you can find vestiges

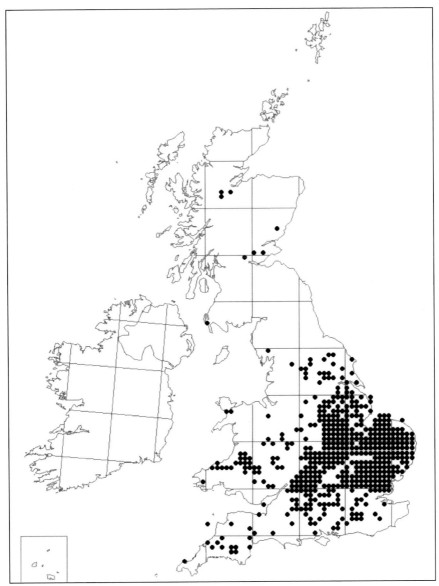

Muntjac Distribution in the UK from the British Deer Society's 2001 Survey and reproduced by kind permission.

of the second and fifth metacarpal bones which in their distant ancestors articulated with them.

The legs themselves are extremely slender, hardly strong enough,
one would think, to support the muntjac's well-muscled, sturdy body.
In some but not all individuals, one cleave is shorter than the other,

REEVES MUNTJAC (*Muntiacus reevesii*)

Male: buck. Female: doe. Young: fawn.

HEIGHT AT SHOULDER:	43–46cm (17–18m).
WEIGHT:	13–15kg (30–35lb). New-born fawn approx 1kg (2¼lb).
PELAGE:	Summer: bright chestnut; chin, throat and underside of tail white; no spots. Winter: deep brown to grey. Prominent ears, translucent in sunshine. Long tail, 15cm (6in), carried erect in panic flight. Surrounding white hairs erectile. Fawns: spotted for first eight weeks of life; spots light coloured, in lines; clear strip of chestnut down the back. Bucks have projecting upper canines up to 2.5cm (1–1½in) long; females 0.5cm (¼in).
ANTLERS:	Males bear short antlers, up to 15cm (6in) on long pedicles, which continue as ridges down the forehead. Sometimes short brow line, otherwise simple spikes. Antlers shed annually, but time of year varies between individuals.
BREEDING:	Non-seasonal. One fawn produced roughly every seven months. Gestation 210 days. Mating follows shortly after birth. Lactation six to eight weeks.
DISTRIBUTION:	Spreading. Mainly Midlands, north to Derbyshire. Fast Anglia, south to Sussex-Wiltshire. Welsh border. Isolated colonies elsewhere.
HABITAT:	Thick undergrowth.
FOOD PREFERENCES:	Mixed grazers and browsers. Principally ivy, bramble, grass. Also fruit, nuts, dead leaves, fungi, market garden produce.
HABITS:	Territorial. Social unit: family group. Two pairs of head glands used for scenting. Fraying mainly with incisors. Need for water in dry weather. Barks freely at intruders, often for many minutes. Young mostly ejected before arrival of next fawn. First colonists are young males. Faeces often deposited in heaps.
VOICE:	A harsh persistent yapping bark often continued for many minutes. Both sexes. A clicking noise is sometimes made when running. Adults and young squeak, sometimes prolonged to a series of rapid squeals. Fawns may call with bird-like piping.
IDENTIFICATION:	Distinguish from roe or Chinese water deer by typical hunched stance. Long tail, short legs. Long pedicles. Prominent frontal and suborbital glands. Short tusks.

giving an uneven footmark or *slot*. One of the best ways of looking for the presence of muntjac is to search for slots in soft or muddy places in the woods. They seem to run the length of rides more than other deer, rather than crossing them. One will also start to notice their tunnels in the brambles, which will be well-trodden. The slots of an adult muntjac are about 20mm (¾in) long by 13 or 14mm (½in) wide. The hind feet are slightly shorter and wider than the forefeet. Where a single animal has left clear tracks you can sometimes measure the stride, which at the walk is about 28cm (11in).

Some claim that muntjac are ugly. Close acquaintance with them brings affection. They are, in fact, engaging little animals with their button noses and lustrous dark eyes, and few people could fail to respond to the miniature charm of a young fawn. Like all deer, fawns are spotted although the spots may not be as well defined as with other species. They soon begin to fade and will have almost disappeared by the age of four or five weeks, by which time the fluffy juvenile coat will be replaced by pelage resembling the adult's.

Muntjac have a wide distribution in southern and south-east Asia, from India to south-east China and Formosa, including Sumatra, Java and some other Indonesian islands. A number of sub-species of muntjac have been proposed. Reeves or Chinese muntjac are found in south-east China. They are named after John Reeves who in 1812 was appointed Assistant Inspector of Tea for the East India Company in Canton. His researches into local natural history were rewarded by Fellowship of the Royal Society and election to the Linnean Society. In addition to having his name linked with the Chinese muntjac, the spectacular Reeves pheasant was also named after him.

The Duke of Bedford was responsible for introducing Reeves Muntjac to Britain some time after 1900 as a replacement for the larger Indian species, which were eliminated inside Woburn Park but were reputed to have survived outside it at least until the 1920s. At one time it was believed that the two races hybridised to produce animals intermediate in size. However, advances in the technique of haemoglobin analysis cast doubt on this theory. Certainly the barking deer which are spreading year by year throughout England either started as pure Reeves or have through successive generations bred out any Indian characteristics. Their escape and spread from Woburn, often with deliberate or unintentional assistance from man, is the most interesting news for the deer watcher since the reintroduction of roe to Dorset around

1800. They have discovered and exploited an unoccupied ecological niche; the woodlands, parks, shrubberies and gardens of suburbia. Deer watching for the town dweller, once a matter of early hours and long journeys, now is found literally on the doorstep.

Muntjac are elusive and shy where they have learnt to mistrust people or their dogs, although in undisturbed woodland one can encounter them wandering about at any time of the day, usually in fairly thick cover. Like any ruminant they prefer to feed for a short while and then lie down to chew the cud. As my own home is only on the edge of muntjac country, my acquaintance with them is not really intimate though we are seeing the arrival of occasional colonists. Enthusiasts who have studied them closely, speak of 'lairs' or hiding places under thick cover to which they return. One can find these sheltered hollows which do look as if they are used habitually, something which is not typical of other deer.

In her book *Muntjac* Eileen Soper gave an interesting list not only of the wild food which she observed her semi-tame muntjac to take, but also of items which she supplied, some of which were accepted and some rejected. The list of wild food included buttercups, hog-weed, dandelions, ivy, dock, grass and holly, various fungi, and wild fruit and berries including crab-apples, rosehips, rowan-berries, sweet chestnuts, horse-chestnuts and acorns. Of the food she put out, apples, pears, peanuts, acorns, sweet chestnuts, horse-chestnuts and walnuts were accepted, but carrots, turnips, parsnips, celery, brown bread, oats, flaked maize, sultanas, potatoes, green peas, beans, bananas, raspberries, gooseberries, tinned pineapple and tinned peaches were refused. This list is, no doubt, not comprehensive but can give a good idea of the muntjac's food preferences and also what might tempt them out into places where they can be observed. Gamekeepers confirm their liking for the feed put out for pheasants. The *Handbook of British Mammals* says: 'shrubs, grass roots, little damage to trees, but visits farm crops and orchards and is known to damage brassicas and roots'.

The Forestry Commission, however, has expressed concern at the damage muntjac are doing to their young trees in the eastern counties. According to their mammal research team, fraying damage is minimal but browsing damage is significant in some places. Muntjac are known to rear up on their hind legs in order to get at some tit-bit but they also walk over small trees, bending them down to eat the tops so that the damage is much higher than one would normally attribute to such a

Muntjac are small, well-adapted to life in suburbia. Photo Brian Phipps

small deer. It is now clear that they are a concern to conservationists because of their liking for bluebell and orchid flowers and seeds. In addition the present-day drive to re-establish coppice can be inhibited if muntjac are present in high numbers.

With muntjac in particular there are plenty of fascinating topics to speculate about. For example, are we seeing the gradual adaptation to our temperate climate of a semi-tropical animal, and therefore one which has no seasonal timetable? There are indications that muntjac may already be in a state of change. They do already grow a winter coat which is complete in November, and antler development takes place annually from May to September. A buck born early in the year may have developed his pedicles sufficiently to grow his first antlers by the age of eight or nine months, while one born in March or later may not achieve this until the age of fourteen months, not being in hard horn with his first head until the age of a year and a half. Normally the antlers are single spikes, in older bucks hooked at the end. Probably in response to good feeding, some bucks develop a short brow point.

The female breeding cycle remains firmly non-seasonal. The fawns are born at any time of year and weigh about 1kg (2¼lb) at birth. They very quickly become mobile and put on weight rapidly, gaining about 0.5kg (1lb) a week while they are dependent on the doe. From the first mating a doe in the wild can be expected to be continuously pregnant for the rest of her breeding life, as mating occurs between eighteen hours to three days after the birth of the fawn. Courtship is signalled by long spells of barking, and hectic chases in the undergrowth.

The scenting habits of muntjac are difficult for us to comprehend, although it is obvious from their behaviour as well as the battery of scent glands with which they are equipped that scent is as important to muntjac as it is to other deer species, or even more so. As well as their frontal and suborbital paired glands, they also have glands between the cleaves of their hind feet and associated with the urino-genital system of both male and female. The glands in the foot are obviously useful in allowing a family which has become separated to follow one another in thick cover.

One apparent effect of possessing scent glands in the reproductive tract is to make both sexes mutually attractive more or less throughout pregnancy. Buck and doe appear to welcome a degree of contact which is unthinkable in other deer except at the height of the rut. The pair often indulge in mutual grooming and licking. Enthusiastic chases are

A muntjac buck in early velvet. Their primitive origins can be seen from the possession of tusks in the upper jaw and prominent glands below the eyes and between the ridges of the pedicles. Photo Brian Phipps

also recorded throughout the breeding cycle. In general, there appears to he an unusually strong pair bond, the nature and extent of which cries out for study.

The droppings which are roughly ovoid, about 2cm (1¾in) long and usually faceted, tend to be deposited in heaps. Several deer may defecate or urinate at the same place and the opening and closing of the suborbital and frontal glands at such times indicate that these functions have a social significance. The bucks set scent from these glands on to their droppings and their food as well as on fraying stocks scattered round the territory.

All observations so far suggest that muntjac are as fiercely territorial as roe deer, but there are indications that under certain circumstances, possibly higher densities, a degree of tolerance exists between mother and young beyond the birth of the next fawn. If this tolerance is connected with density, then one could expect a threshold below which the young are automatically ejected once they are independent; but if through complete colonisation and insufficient culling the population density passes this point, tolerance may allow a much higher number to exist in the same area. This indeed does seem to be the case in some of the woods in Northamptonshire for example.

Barking is one of the first ways in which the presence of muntjac may be suspected. They yap like a small dog and, unlike the roe, may keep it up for many minutes on end. Is it aggression, curiosity, love or hate? Are there different barks for different occasions? These questions can only be tackled with the insight you get from long hours in the woods.

Muntjac are known to have a wide vocabulary including the high squeal of a doe in season and the shrill pipe of the fawn. A buck in pursuit of a female repeats a curious little grunt, and they will grind their molars and chatter if annoyed. A clicking noise is sometimes made, usually, but not invariably in flight. In terror, muntjac have a distressing loud scream.

A muntjac doe in summer coat. Note the long tail which distinguishes them from roe.
Photo Brian Phipps

When they have been ejected by their parents, young deer wander widely and in doing so risk death or injury on the roads. Statistics of road accidents show that sixty per cent of the muntjac involved are males, which probably indicates that they wander more widely or are chased out in a more determined fashion by their parents than are young does.

It is noticeable on the outskirts of muntjac colonised ground that the first to be recorded are almost invariably bucks. The establishment of a breeding population may not follow for several years after the first records of intrusive males. Even so, they are consolidating their hold on the country at a rate which has colonised the East of England north and west to a line from the Humber to Dorset. This area carries the heaviest human population with, one would have thought, little scope in the urban sprawl for colonisation by any animal larger than a rabbit. The muntjac proves this wrong. At present, neither climate nor human population density seems to limit their spread. Isolated sightings come in from an even wider area.

While there may be larger or more graceful deer to watch in more spectacular surroundings in other parts of the country, none of them offers a greater challenge to the enthusiast either from pure technical difficulty or from the rewards to be gained from observing an animal about which there is much to learn. At the moment, who would be brave enough even to say whether or not muntjac really are our most numerous deer?

8 • Roe – The Fairy of the Woods

In the tales of childhood the fairies always seemed rather improbable insect-like creatures, hovering around flowers and so on. By contrast, brownies were sturdy little people living on their wits and their knowledge of lumbering humankind, always ready to slip behind the nearest oak tree before continuing with their secret revels. Could a roebuck have been the original inspiration for the brownie, living pucklike as he does in our very gardens and often totally unsuspected?

Roe are small, beautiful and brown, as active and mischievous as any woodland sprite. They may be detested by some people for the damage they do, but even those who suffer most from their attentions have a sneaking respect for the cheeky way they live at our expense. Like Puck himself, roe have lived in Britain a very long time. They were here when London was a steaming, hippo-haunted swamp. They saw the inexorable swing of climate which ushered in the last cold phase of what we know as the Ice Age, and they survived when other forms disappeared; the hyena, the bear, the giant deer and the woolly mammoth.

Their bones have been found in the rubbish dumps of Roman army camps. Saxon and Norman protected them and roe were included in the laws against poaching passed by Richard Lionheart for which the penalty was castration. From the fourteenth century the tide of popularity turned against roe because of an idea that they drove away other deer which were considered more worthy quarry. Demoted in the hierarchy of legal protection from equality with red deer and wild boar as Beasts of the Forest, they were thereafter merely Beasts of Warren, so that poaching them no longer made one liable to the terrible severity of the Forest Law. Once plentiful throughout the country, roe became progressively more and more scarce in the Midlands and South of England so that writers of the eighteenth century spoke of them as 'formerly abundant', while even in Scotland they retreated into the recesses of the Highlands.

The fact that roe did not go the way of the reindeer and the elk is due partly to their charm, which led to reintroductions, and partly to their innate capacity to survive and adapt to changing conditions. Generally speaking, by the beginning of the nineteenth century they were only to be found in the north of the Highland line. By 1900 the whole of Scotland to the Border country had been reoccupied. What changes in the environment could have taken place which were profound enough to have been such an enormous benefit to the roe?

During this time, of course, Scotland had seen the full effect of the Highland clearances, when the glens were cleared for sheep of whole communities who, before their eviction would have been glad enough to stave off hunger by snaring a roe. Sheep farming itself soon failed, and many Highland proprietors were forced to let or sell their properties to the new rich of the south to whom game and game preservation was paramount. No doubt the roe of the lower ground birchwoods benefited as much as their larger cousins the red deer of the open hill. Another factor was that tree planting came into vogue in the wake of the Napoleonic wars. Nothing favours roe deer more than the high living and safe lying of a young plantation.

Even so, these are local factors which are not sufficient to explain a reversal in the fortunes of roe which has led to a surge in numbers and an increase in colonisation throughout Europe. They are still on the increase. The answer could be social or political, the same influences which contributed to a population explosion of elk in Sweden. I was told that before the war Swedish farmers used to pasture their cattle in

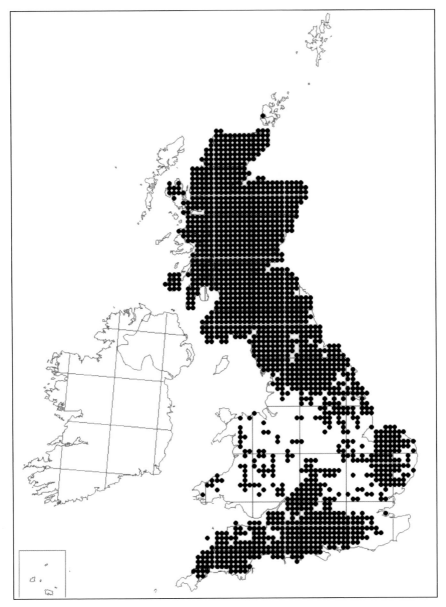

Roe Deer Distribution in the UK from the British Deer Society's 2001 Survey and reproduced by kind permission.

the woods during the summer. Presumably they were looked after by small boys who are now expected to be in school, so the practice has died out, leaving more browse for the elk. After World War Two the exploitation of timber was stepped up with a corresponding increase in

ROE DEER (*Capreolus capreolus*)

Male: buck. Female: doe. Young: kid or fawn.

LENGTH:	120cm (47in).
HEIGHT AT SHOULDER:	65–73cm (26–29in).
WEIGHT:	23–32kg (50–70lb) live weight.
PELAGE:	Summer: bright red-brown, pale below; no spots; rump patch lemon yellow or buff to white; nose black; chin white; no visible tail. Winter: dark brown to grey; thicker; rump white, hairs erectile on alarm; sometimes white patches on throat. Fawn: new-born has splashed pattern of irregular white spots on variegated ground of brown and black; spots fade August–September.
ANTLERS:	Typically six point. Shed October–December. In velvet to March–April.
BREEDING:	Rut July–August. Fawns born May (delayed implantation). Usually twins. Triplets not uncommon but rarely survive.
DISTRIBUTION:	World: Palaearctic region, absent N America, NW India, N Africa. UK: generally distributed Scotland and N England to a line Humber–Mersey. Welsh border (local), southern counties, W Kent to E Cornwall and north to Buckinghamshire, Gloucestershire. Some islands eg Islay, Skye, Bute.
HABITAT:	Woodland, preferably with open patches. Moorland.
FOOD PREFERENCES:	Leaves, buds and twigs of most trees and shrubs. Some grass, cereals and herbs. Occasionally fungi and fruit.

HABITS: Solitary, or family groups. Fiercely territorial. Threat displays of scraping, fraying, barking. Most young ejected annually. Not nocturnal, but adapts feeding hours to avoid disturbance.

VOICE: Both sexes bark like a small dog in alarm or challenge. Barking may continue for up to a minute but not prolonged like muntjac. Bucks grunt and wheeze in rut. Does have high-pitched squeak to attract buck. Juveniles may squeal loudly in fright and communicate among the group by a nearly inaudible bleat.

IDENTIFICATION: Distinguish from muntjac by upright stance, long pointed ears, no visible tail. Chinese water deer has prominent canines, short dark tail; males antlerless.

second growth for the deer to enjoy. Certainly it has been noticed that as the elk have pushed south, so the roe in Sweden are pushing north towards the Arctic circle.

The underlying factor may, on the other hand, be climatic. We know that relatively small shifts in the flow of the Gulf Stream can have a profound effect on the weather. Between the fifteenth and the nineteenth centuries, Northern Europe experienced a drop in average temperature which was known as the Little Ice Age. Maybe it is significant that this period also saw a low point in the fortunes of roe from which they are now recovering. Possibly the process is accelerated by the effects of global warming. Speculations of this sort do not prove anything, but can be interesting and even useful in understanding the background to deer behaviour.

The story of the reintroduction of roe to southern England has often been told. Love of hunting led Lord Dorchester around 1800 to establish the first colony at Milton Abbas in Dorset, from which most of the roe from Hampshire westwards are descended. The original animals

A mature roebuck in summer with his typical six-point antlers, which can be up to 30cm long.

may have come from France. Roe from Petworth Park in Sussex now occupy the south-east of the country, from the Kentish border to central Hampshire and north to the Chilterns. Why they have been slow to penetrate into Kent is something of a mystery. The ancestors of the roe of Norfolk and Suffolk were brought from Germany to Thetford in 1883. They were quick to take advantage when the Forestry Commission started planting up the Norfolk Breckland after World War One. For the next forty years or so they were rarely to be found outside the boundaries of Thetford Chase, but when colonisation of this vast area was complete roe soon began to be seen on surrounding farmland. They are now firmly established over much of East Anglia and are spreading into the East Midlands.

While these reintroductions were taking place, an invasion of native Scottish roe started to penetrate Northumberland and Cumbria. Once again they rapidly took advantage of the Forestry Commission's zeal in creating new cover for them. From Kielder to Newcastleton the sprawling new woodlands were occupied as soon as there were trees tall enough to hide a roe, creating unimagined problems for the future.

The latest British Deer Society Survey shows near-complete colonisation by roe of the North and East of England from Essex to Cumbria, solid establishment from West Kent to Cornwall and from Wiltshire north to the Central Midlands and into Wales. Scotland has roe in every county.

The roving deer watcher will be able to see small differences between one local race and another. For example, all the year Scottish roe have a very white tail patch or 'target' (roe do not have a visible tail) while among the south country roe it is usually lemon yellow or even brownish in the summer. The ear on the northern race is slightly shorter in proportion to the skull and there are small differences in the type of the antlers grown, although this may only be obvious when one looks at a collection of skulls from one area. There is a lot of variation between individual antlers. Bucks from Sussex tend to grow antlers which are thick and heavily pearled but lacking beauty to some extent because the beams are rather near together. West Country heads, in contrast, often have a wide span with long points but are more delicate in build. The best Scottish heads have a classical elegance about them; a fine balance between the back points, which are often long and elegantly curved, and the forward-jutting brows. One can expect these to be set a little higher on the beam than is common with roe antlers from the south.

A roe doe in summer coat. Although roe have no visible tail, in winter coat does grow a downward-pointing tuft of hair in the centre of a prominent white backside.

The result of all this expansion of the roe population is good news for the deer watcher. There are now few places which are more than an hour's drive from roe-occupied country. Not so good for the forester, who sees his planting policy threatened by the possibility of severe damage. The measures which he is forced to take in order to protect his trees are expensive and time-consuming. For example, broadleaved species may have to be planted in plastic tubes, or whole plantations may have to be deer fenced. To the casual eye, a few deer-damaged trees may not look serious, but if the damage passes a certain critical point then patches may have to be replanted – 'beaten up' in forestry terms. This is not only costly but puts off the stage when the young trees start to compete successfully with the surrounding vegetation and can be left to grow on without further expensive weeding.

Although there are now a number of sophisticated ways of reducing the damage which deer do to young plantations, it is still generally true that if there are too many roe about, they will do quite unacceptable damage until the young trees are tall enough for the leading shoots to be out of reach. Depending on the sort of tree that is planted, the worst damage will be within the first four or six years. One could say 'if that is so, why do you have to go on shooting roe year after year?' One of the answers to this is that forestry is a continuous process. Even in the same wood one patch may be growing up while not far away another plot is mature, due for clear felling and replanting. No wonder foresters regard roe as a serious pest and employ professional stalkers or rangers to control their numbers.

In farming areas many of the deer will be out on the fields in the spring because there is little food for them in the woods. The does will still be accompanied by last year's young, but as they get nearer to fawning once again, their offspring will be ejected to make room for the new family. It is a very traumatic experience for a young buck to be chased hither and thither by older territorial males or for his sister to suddenly find the devoted mother of the last eleven months viciously turning against her. By late May or early June yearling buck and doe alike are driven away. They may wander great distances, until they find either unoccupied country or some corner where food and cover is too

New-born roe kids lie still for long periods until rejoined by their mother. Spots and slashes of light and dark make them difficult to spot. If you find one, leave it alone – it has not been abandoned. Photo Brian Phipps

After a few weeks the kids are active enough to follow their mother. The spots fade by early autumn. Photo Brian Phipps

scanty to attract a territorial adult. This may be a new plantation, a scrubby area liable to be invaded by cows or picnic parties, or the inhospitable recesses of a conifer plantation.

A certain number of young animals escape this exodus. Occasionally a doe fawn will be tolerated so that she sets up a satellite territory adjacent to her mother's, while some yearling bucks, usually those that were lightest in weight at birth, are so unaggressive that they are considered unworthy of the territorial buck's attention and live a furtive life, virtually fagging to the older beast. Research has shown that these have a better chance than their more robust brothers of eventually achieving and holding a territory themselves.

Spring is a fascinating time to be out watching the deer about their daily affairs as the woods green over and the dawn chorus reaches a crescendo. If you ever have the offer of a summer morning in the company of a deer enthusiast, never pass up the chance. It is likely to be an unforgettable experience. Make sure that you can find the rendezvous in the dim light before dawn. Wear soft, unobtrusive clothing, and light rubber-soled shoes. Take your binoculars of course, and be very sure not to be late. You will notice that the stalker never slams a car door. Noise travels far in the dawn quiet. You will try to imitate his slow, silent progress, looking first at the ground to avoid cracking a stick or rustling the leaves, and then all round to catch the flick of movement or any suspicious patch of colour which may reveal a roe.

You go upwind or across it, or you will not see many deer. By the end of two hours or so you may have walked a mile, or only a couple of hundred metres, but you will be tired. This way of walking has no rhythm in it and the concentration needed between quiet movement and constant spying is very demanding. In the end, increasing sounds of human activity show that the workaday world has come to life and the deer will be about no longer. Maybe you will have been lucky to have spotted several, or none at all, but the way of going about a deer watching session can be learned so much more easily and quickly from an experienced stalker than from any book. Such an outing is, however, a great privilege and even if you know someone who is a professional stalker or an amateur, do not think that the invitation to come out is given lightly.

Breeding cycle

Roe have a different timetable from the other deer, partly due to an oddity of their biology. An increase in barking and the progressive thickening of bucks' necks during July are welcome signs to the deer watcher that the rut is approaching. Towards the end of the month courtship will be in full swing. One can see bucks searching for does, nose to ground, and if you are lucky enough you can witness the hectic chases which often finish with the buck and doe running in a tight circle round some object like a tree stump, or a bush, until a trodden path develops which is called a *roe ring*. Most roe prefer to run anti-clockwise for the same reason that most of us are right-handed. You can check this by looking at the direction in which the grass has been trampled down. The doe has a variety of squeaks which she makes both to attract a buck to her and during courtship. The buck, as well as barking, makes a wheezing noise – once heard never forgotten. When the doe is ready the buck may mount her many times. She may accept more than one buck if the occasion offers.

Once fertilisation has been achieved the cells start to divide until a tiny ball of tissue is formed about the size of a pinhead. In mammals with a normal reproductive system, this *blastocyst* (the fertilised egg) would continue to develop with the establishment of a placental link through which the growing foetus receives its nutriment from the mother and eliminates its waste products. In roe, this process is delayed. The blastocysts lie more or less dormant until the end of the year, growing in this time to a diameter of about 20mm. At the turn of the year implantation in the wall of the uterus finally occurs and development after that proceeds at a normal pace. Why roe should have developed this mechanism is not clear. It could have been an adaptation to a cold phase in the climate when both fawning and rut had to take place within a short sub-arctic summer. Until some well-preserved prehistoric roe emerges from its ice block, like a mammoth, we will probably never know.

Most fawns are born in May, which means that the total length of pregnancy is nearly ten months. Apart from the muntjac, which is a semi-tropical animal and therefore a non-seasonal breeder, other deer do not rut until later in the year so that normal implantation and development brings the time of birth again to the early summer when the mother has the best chance of providing an abundant supply of milk.

When her time comes, the doe is very secretive and she retires into thick cover for the birth of her fawns. This is not always as easy as one imagines for wild animals and, of course, if anything goes wrong or the fawn is badly presented, there is no one to help. On occasion I have found the pathetic corpse of a doe which died fawning, and there can be no sadder sight. One longs to have been on hand to help. Strangely enough, roe seem to be able to sense when help is offered in dire need. Once a doe with a snare round her neck stood stock still while the deadly wire was eased over her head. Twice I have heard of does in labour with the fawn badly presented allowing human help in turning and delivering it without struggle or protest.

Happily most births are successful, though labour can last an hour more, and one fawn may well be dropped at some distance from the second. It is likely that the first comer will not have been licked dry by the mother and if the weather is cold and wet it will be at risk from chilling. Many die in years when the weather is bad towards the end of May.

Before the fawns are strong enough to follow her, the doe hides them. When she goes out to feed they have a strong instinct to lie doggo. They are very charming little creatures. If you stumble across one it will look so beautiful and confiding that the urge to stroke it or pick it up is almost irresistible. So many people who find a fawn fail to realise that it has merely been put to bed for an hour or two and the mother will be close by. Every year I have a spate of telephone calls from kindhearted folk who have discovered a 'Bambi' and assuming it to be abandoned have taken it home. From the fawn's point of view there could be no greater disaster. The mother is distressed, deprived of her young and for anyone to try to bring fawns up by hand leads only to heartbreak in the end. They are difficult to rear and, even in the rare event of success, the average garden is totally unsuitable as habitat for a full grown roe, let alone the way it will devastate any flowers or vegetables which are planted.

It is impossible to return a hand-tame animal to the wild and no wildlife park will be interested in giving it a home. In addition, should the fawn be male, by the age of a year it is likely to be really dangerous. Having no fear of man, a roebuck can turn on those it knows best in the fever of the rut and inflict serious or even fatal injuries. Tame roebucks have been responsible for human deaths in the past.

So if you find a fawn, *leave it where it is*. Do not disturb the grass or herbs round it, and having admired it from a distance, go away. If it is in a field and there is a risk of it being killed with grass-cutting machinery, carry the fawn to the nearest hedge and leave it there. After

Roe deer in winter coat. Photo Brian Phipps

a while it will start squeaking and the doe will come to find it. Even after this handling it will have a very much better chance of survival than by any attempt to rear it artificially.

Antlers and feeding

Again in contrast to the other deer species, roe cast their antlers in the autumn and are in velvet during the winter and early spring. Antler growth is a drain on the resources of an animal and although calcium and other minerals can be stored and withdrawn at need from the skeletal structure, principally the long bones, protein is needed to form the basic structure of the antlers and this cannot be stored. A daily intake through the winter is necessary. For this reason, the size of antlers grown by a buck varies according to the protein resources available. Depending on how and where he winters, an individual buck may grow large antlers one year and smaller ones the next regardless of age. This makes roe impossible to identify from one year to another by their antlers, a frequent source of misunderstanding and difficulty among those who try to manage them.

It is known that roe, and probably other deer as well, can regulate their metabolism to some extent so that during rest periods in the year and in bad weather they can eat little and yet not suffer great loss of bodyweight, providing they do not have to move much. Bucks are very hard to find for a month after the rut and all roe are elusive in November and December to the extent that one imagines they have all moved away or been poached. After the turn of the year, however, when the does start to become visibly pregnant, and the bucks are territory-minded once more, they reappear. Areas which seemed devoid of deer can suddenly be busy, with little family groups to be seen in the early morning and late evening feeding in the fields or on whatever browse the winter has left for them.

From the deer watcher's point of view roe may be the most rewarding species to study but they are almost certainly the most difficult. Because they are so clever at adapting their behaviour to different habitats, even long experience of watching roe in one part of the country may not help very much in another. The roe of suburbia use their familiarity with man to dodge like the brownie behind the nearest tree, or even to stand motionless knowing that the majority of walkers will fail to see them. In a very different setting, the roe that live in the remote forests and heather uplands of the north who see few men in the course of the year could be expected to be naive and confiding. Far from it. One glimpse of distant movement and they are off, sometimes travelling a long way before settling again. Roe on the open hill have to be stalked with infinite care and patience, just like red deer, with pains being taken not only with the wind so that they do not scent you but to keep out of sight as well. Otherwise they will be off when you are still a long way away. Success in approaching them, perhaps close enough for a photograph, is an achievement. Even a passing acquaintance with roe in any surroundings is a wonderful experience.

9 • Red Deer – Monarch of the Glen

Meet a red stag face to face and you don't need to be told that it is our largest land mammal. It is huge; impressive or frightening depending on the circumstances of the meeting, and somehow very appropriate to its setting, whether this happens to be an English wood or a Scottish mountainside. Legends have been woven round the stag, from Herne the Hunter to Robin Hood. Even the age to which he would live was a mystery:

> Thrice the age of a dog, the age of a horse.
> Thrice the age of a horse, the age of a man.
> Thrice the age of a man, the age of a stag.
> Thrice the age of a stag, the age of an eagle.

Yet in fact a twenty-year-old stag is very old indeed.

The Neolithic flint miners of four thousand years ago used stags' antlers as picks, and no doubt they had many other uses for antlers and bones, besides enjoying the meat. Red deer by nature are animals of the

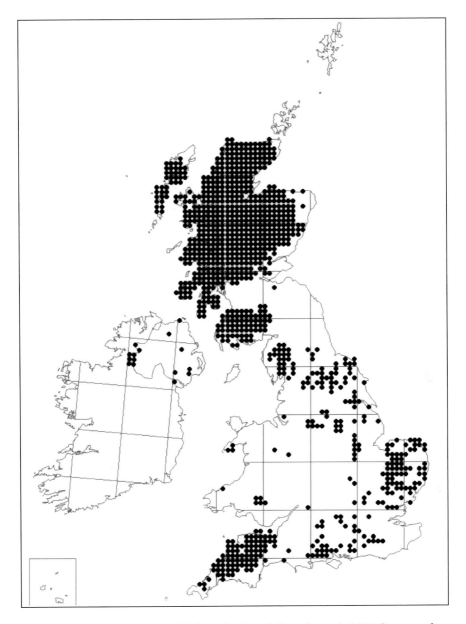

Red Deer Distribution in the UK from the British Deer Society's 2001 Survey and reproduced by kind permission.

forest and the forest edge. The heavily wooded landscape of post-glacial Britain suited them. To judge by the numerous remains which have been dug up, they were abundant and grew to a very large size, possibly

RED DEER (*Cervus elaphus*)

Male: stag. Female: hind. Young: calf. Juvenile male: pricket, knobber.

LENGTH:	150–190cm (60–76in).
HEIGHT AT SHOULDER:	105–140cm (41–54in).
WEIGHT:	Variable according to habitat and feeding. Male: 90–190kg (200–420lb) or more. Female: 57–115kg (125–250lb) live weights.
PELAGE :	Summer: dark red to brown; rump and inner thighs cream or yellow; no spots. Odd individuals, and some strains of park deer, have a pattern of pale spots, especially along the back, in summer coat. Tail length *c* 15cm (6in). Stag develops mane before the rut. Winter: dark brown to grey, rough. Calf: brown thickly spotted with white.
ANTLERS:	Typically branched, multi-pointed, round in section. Number of points no indication of age. Shed March–April (juvenile sometimes much later). In velvet to August–September. Antlerlers stags (rare) called 'hummels' (Scotland) or 'nott' stags (W England); can breed.
BREEDING:	Rut September–October. Calves born end May–June. Twins very rare.
DISTRIBUTION:	World: Palaearctic region to Manchuria, S Himalayas. Many sub-species. Introduced S America, New Zealand. UK: Scottish Highlands and Islands, Galloway, Lake District, SW England (mainly Quantocks, Exmoor and W Devon). Many other small colonies, mostly due to park escapes. Present on many deer farms.

HABITAT:	Typically open forest. Has adapted in Scotland to living in open, on high hills in summer, moving to lower ground in winter.
FOOD PREFERENCES:	Most agricultural crops, heather, leaves and twigs of trees and shrubs. Seaweed on coasts. Fruit and fungi.
HABITS:	Separate herds of stags or hinds and juveniles most of year. Smaller groups in woodland. In rut, especially on open hill, stags gather large hind groups, defending them from other males and roaring throughout the day. Both sexes wallow. Calf and yearling tend to stay with hind. May hybridise with sika.
VOICE:	Hinds communicate with their calves by a muted lowing call. They also bellow loudly while calving. Alarm call is a loud, gruff bark repeated at long intervals. Calves communicate with their mothers by a high-pitched bleat. In the rut the stag roars; a variety of loud noises from a long-drawn cow-like moo to a series of staccato coughing grunts.
IDENTIFICATION:	Large size. Short tail. No spots. Sika (and most hybrids) have white mark below hock. Fallow have flattened antlers, long tail, and are spotted in summer.

only approached these days by stags from eastern Europe. Some magnificent sets of antlers have been recovered from the peat in Scotland to show what this animal is capable of achieving, given an abundant supply of food and a kinder climate.

As the human population grew, the wildwood began to be cleared, bringing the deer under pressure. Their fortunes north and south of the border, however, took different ways. In England, sport became the ruling passion of the aristocracy from the earliest times, the excuse being that it was the finest training for war. Even as early as the seventh century AD there is evidence of the existence of hunting reserves in Europe, and by the time of King Canute (reigned 1016-35) laws regulating hunting and reserving it for the sovereign were already in force.

Deer were given an increasing degree of protection by the Normans, especially in the royal forests where savage sentences on commoners were imposed for offences against 'venison or vert' – the king's deer and the trees which fed them.

In Scotland, by contrast, the deer for a long time were just regarded as a food resource. By the dawn of history much of the broadleaved woodland which once covered much of Scotland had been replaced by Scots pine to form the vast Caledonian forest. Not only red deer, but wolves and outlaws lurked in its recesses. One technique which was widely used for getting rid of these pests was simply to burn the forests, and whole mountainsides were put to the torch. What remained began to be used up for iron smelting and domestic fuel.

Because the deer were not protected to provide hunting, their numbers declined with the habitat and they had to fall back on that saving grace of the deer family – adaptability. In complete contrast to their heredity they took to the open hillsides and mountain tops. At first, no doubt, they would move up and out for the summer, sharing the winter grazing of the lower ground with the Highlanders and their cattle. Later they had to compete with sheep which drove them farther into the hills, so that by the time Queen Victoria came to the throne, deer were pretty scarce even in the Highlands. Once again, it was hunting, though of a very different nature, that was to prove their salvation.

The deer, in fact, would probably not be there today but for the rise in popularity of the sport of deer stalking in the nineteenth century. No matter what one's feelings may be about shooting, red deer do now represent a major asset which in remote areas may literally be the only source of revenue and employment. The possibility of conflict does exist between those whose job or whose sport it is to take the necessary harvest of surplus deer each year and those who would merely like to watch them or to enjoy the freedom of the hills in other ways. So it is worth knowing a little about how stalking came to be such an important part of the Highland economy.

Deer stalking – ancient and modern

The vogue for deer stalking as a sport started early in the nineteenth century. The Highlands had only recently endured two traumatic upheavals. Before the rebellion of 1745 the high hills were inhabited, at least during the summer, by a numerous if penurious community of clansmen dependent on their local laird, who would no more have gone out to kill a deer for the larder than he would have dug potatoes. It was not consistent with his dignity. During the Middle Ages there were, however, great communal hunts known as *tainchel*, sometimes in the presence of the sovereign. An army of clansmen would spend weeks combing enormous areas for deer which were driven into an enclosed space and there butchered. Some idea of the scope of these massacres can be gathered from an account of a tainchel arranged for Queen Mary in 1563 in the forests of Mar and Atholl:

> The Earl of Atholl prepared for Her Majesty's reception by sending out about two thousand Highlanders to gather the deer from Mar, Badenoch, Moray and Atholl to the district he had previously appointed. It occupied the Highlanders for several weeks in driving the deer, to the amount of two thousand, besides roes, does, and other game . . . Her Majesty having ordered a large fierce dog to be let loose on a wolf that appeared, the leading deer were terrified at the sight of the dog, turned back and began to fly whence they had come. All the other deer instantly followed. They were surrounded on that side by a line of Highlanders but well did they know the power of this close phalanx of deer and at speed, and therefore they yielded and opposed no resistance and the only means left of saving their lives was to fall flat on the heath in the best posture they could and allow the deer to run over them. In this manner the deer would have all escaped had not the huntsman, accustomed to such events, gone after them and with great dexterity headed and turned a detachment in the rear. Against these the Queen's staghounds and those of the nobility were loosed and a successful chase ensued. Three hundred and sixty deer were killed, five wolves and some roes, and the Queen and her party returned to Blair delighted with the sport.

Even if, as is possible, this story is highly embroidered, there is no doubt that great drives were made and that dogs, bows and other weapons were used to dispatch the deer that had been encircled.

After the rebellion of 1745 the clan system was swept away in a wave of retribution and sequestration. At the same time large areas of the Highlands began to be cleared for sheep, not only of cattle and deer, but of their human inhabitants. The sad ruins of abandoned crofts and shielings can still be seen lining the coast and dotting the floors of the remotest glens. The hard feelings engendered by these enforced evictions linger to this day. While sheep numbers were high the deer declined. There is little doubt that the heavy grazing, and intensive management of the heather sward by repeated burning, drew on the meagre resources of stored fertility built up during the comparatively short time since the Highlands were last scoured by

A fine red stag roars his challenge. Photo Brian Phipps

A master stag may attract many hinds to his harem. Photo Brian Phipps

glaciers. In areas of high rainfall or where the sub-soil is impermeable and acidic, the heather has been replaced by an impoverished sward on which the deer alone seem able to subsist. By a fortunate combination of circumstances the decline in profitability of sheep farming came at a time when the Industrial Revolution had brought great riches to the Midlands and the north of England. Victorian industrial magnates in search of wild sport began to penetrate the remoter areas of Scotland, at first by ship and yacht, and later on by rail as the network was gradually extended.

Deer stalking was supposed to have been launched as a fashionable sport by the publication in 1820 of *The Art of Deer Stalking* which was written by an Englishman, Henry Scrope, describing his adventures among the deer in the forests of Atholl. Edition followed edition throughout the nineteenth century.

The deer at first were scarce but soon the new owners of these vast mountainous areas were vying with one another to build up their stock. Fierce competition developed, not only for the number of stags shot each year, but also the greatest average weight and, of course, for the year's largest trophies.

Regiments of men were put to work carving out pony paths so that the sportsmen could get out to the farthest parts of the forest and the deer could be brought home in triumph on a pack-saddle. Queen Victoria and the Prince Consort became captivated by the Highlands, no doubt loving the tranquillity as much as the scenery as a relief from the stuffy formality of life at Court. Prince Albert was also a very keen shot. He bought Balmoral Estate in 1852 and this set the royal seal of approval on the fashion.

All the financial power and mechanical ingenuity of the era was brought to bear on constructing ever larger and more comfortable mansions in the recesses of the Highlands. To maintain these a small army of indoor and outdoor staff were required. During the latter half of the century and up to the cataclysm of World War One, money was poured into the Highlands at a time when it was desperately required. A tradition of deer management was built up which, although strange to modern ideas, ensured that a breeding herd was maintained for the future. Put simply, the Victorian sportsman would choose the largest

stag that he was allowed to shoot. This decision lay more with his status as a guest than with the possible detriment to the stock by shooting the best specimens. A print of Landseer's *Monarch of the Glen* hung in a million English homes. The summit of the visiting sportsman's ambition was to shoot a twelve pointer, or 'royal'. Only the number of stags which could be shot each year was limited. Little other attempt was made, apart from artificial feeding, to maintain the deer herd. The rest was left to nature which meant, in effect, that many starved to death, but not until the winter when the proprietors and their guests had returned south. 'Out of sight, out of mind.'

With the stringencies and heavy taxation which sprang from World War One, the great houses fell upon lean times. Deer stalking, so long the most soughtafter invitation and the exclusive privilege of the owner and his guests, began to be available through letting to a wider public. Many forests, however, remained in private hands and while the style of living and the number of attendants was progressively reduced, stalking continued in much the same manner as in the Victorian heyday. The guests would go out with a stalker and the number of ponies needed to bring home one stag each. The ghillies, or ponymen, often had to set off well before dawn and walk long miles to some distant rendezvous

where they would be expected to wait, regardless of the weather, until required. On many deer forests one can see shelters of piled rocks which the ghillies built themselves over the years for some protection from driving rain and biting cold.

These days the deer are managed, not only to provide stalking which is usually let either by the day or by the week, but to provide a reasonable crop of venison. Instead of letting them die on the hill for lack of food and shelter, attempts are made to keep the herds within the food resources available, which means that every attempt has to be made to remove at least sixteen per cent of the population every year to keep numbers stable.

From late September until the end of the stalking season on 20 October the rut will be gathering momentum and the corries will echo with the roars of lovesick stags. Because of their habit of covering themselves with mud they will look black and enormous in comparison to the hinds. Unlike the Victorians, who would always choose the largest stag, the present-day rifle will only feel justified in shooting a stag which is poor breeding stock. The deer stalker in the Highlands, whether he is a professional or a visitor, is trying to play his part in the proper management of the herd

Deer stalking and the rambler

The population of red deer in Scotland is something over a quarter of a million. Because of the harsh conditions and meagre food on which they live, the hinds may not have their first calf until they are three or four years old; even after that they have occasional years when they fail to breed. The expected average calf crop is accordingly a meagre sixteen per cent of the population. Even so, that represents 40,000 calves every year from a population of 250,000. By definition, as many individuals must be removed from a population as are born into if it is to remain stable. The fact has to be faced squarely that food is the limiting factor for Highland red deer. If enough of the surplus are not shot each year, the rest will die of starvation. Indeed many do. Although enough stags are shot, probably more than enough, the hind cull is inadequate. The shortfall is made up by a grim harvest reaped during the cold, wet, miserable days of the long northern spring.

The time that the proper cull can be taken is limited: from August to mid-October for stags, although in effect it has to be a much shorter time because of their unapproachability early in the season. Many stags join up into large herds which are difficult to get within range until they start to bicker with one another and disperse at the onset of the rut. Hinds are culled from mid-October to January, with the stalker having to battle against short winter days and long periods of bad weather.

Especially during the stag season there is bound to be some conflict of interest between the stalker and others such as climbers and walkers whose pleasure is in the high tops. People who do not know about deer find it difficult to believe how wild they are when, for lack of a trained eye, they may rarely see a deer at all. Certainly few ramblers realise that a puff of tainted wind may clear a whole corrie and not only ruin the stalker's chances for the day, but keep the deer in such a protracted state of alarm that they are unable to build up the food reserves which are vital to them for survival in the winter.

The would-be deer watcher will, of course, take pains to avoid disturbing the deer, not only because of sympathy for them but for the pleasure of keeping them in sight rather than waving a last farewell as

Hinds are good mothers and may continue suckling their calves until late in the year.
Photo Brian Phipps

they cross over a distant skyline. To find where the stalker lives is easy in a country district. Approached for advice about the best chances of seeing deer, most stalkers will go out of their way to be helpful. Let them show you on your map where to go, and stick to it. Take binoculars, or preferably a telescope, extra warm clothing and a waterproof which is not a blue or orange anorak. Make sure that you have with you a compass, map, food, and a whistle to attract attention in the event of an accident. The hills may look innocent, but they are mountains for all that. Wear boots or shoes with a good pattern on the soles, not wellies, and take a stout stick which will help to steady the telescope or binoculars, employ as a probe for boggy ground, and act as a brake going downhill.

Spend more time sitting down and spying than walking; the deer are difficult to spot and may be much smaller than you expect against the grandeur of the slopes. Having located a group, do not immediately set off in their direction but try to discover any deer that may be lurking in between. Above all, do not try to approach any deer too close. They will be away, probably taking many others with them which with a little more patience would have provided a fascinating day's deer watching.

Return to the forest

What of the future? Owing to the efforts of the Forestry Commission, investment companies and private landowners, Scotland now has nearly 800,000 hectares (2,000,000 acres) of productive forestry, much of it planted on traditional deer wintering ground which is vital for their survival. They have been excluded by hundreds of miles of deer fence. This has meant additional privation and the deer make determined and often successful efforts to break in again, especially in bad weather. Large numbers of starving deer in a young plantation will inevitably damage or destroy many trees. This has resulted in a sad and bitter warfare in which deer are shot regardless of sex and season. When the trees are young, deer can be seen. After a few years they have grown high enough to hide a deer, which will only be encountered by chance unless the forest has been designed with the long-term presence of deer in mind.

A forest-dwelling race of red deer has been brought into existence again. Because no fence can be kept deer-proof for even a short length of time, one can expect the majority of Scotland's new forests to have

Red deer living on the open hill in Scotland endure poor feeding and may be forced down to low ground in bad weather.

substantial herds of resident red deer. Deer that take to living in woodlands will benefit enormously, not only from a better food supply but through having shelter from the winds and bitter weather which take such a toll on the energy resources of animals living in the open. If they are not harassed into oblivion, Scotland's stags may soon rival their prehistoric ancestors. One hopes that they will be allowed to re-establish their natural feeding pattern of lying up in the forest during the day and emerging in the evening on to the open hill to feed. If they are, and they are reasonably undisturbed, the forester will suffer much less damage, and the deer will still be visible as a tourist attraction and a valuable resource.

During the winter and spring red deer in Scotland are easy to see because semi-starvation robs them of their caution and forces them down to lower ground in a battle for life. They will be ragged, bony creatures, shadows of their autumn pride, and one is filled with pity and with admiration for their determination to survive. As soon as there is a tinge of green on the barren hillsides they take once more to the high corries, mounting in the summer days to the windy mountain tops to escape swarms of biting flies and midges which can drive deer, and man, nearly mad.

Between the loss of wintering ground through fencing and the hard attitude to deer taken by foresters and some conservationists, Scottish red deer are going through a thin time. However, there are welcome signs that the fact of the inevitable presence of deer in forests is being grudgingly acknowledged. As a consequence forests will eventually be laid out with the necessary greens and open spaces to allow deer to be properly managed as an asset rather than being harried and destroyed as vermin.

The Forestry Commission has already done a great deal towards welcoming the public and at the same time informing them, through information centres, how they may enjoy the forest to the full without spoiling it for others who come after. A visit to the local Forestry Commission office will provide details of forest walks, nature trails, and other facilities for observing wildlife which they have organised in many parts of the country.

The more interest there is in Scotland's marvellous red deer, and the more their role is recognised in contributing to Scotland's prosperity, the more pressure will be exerted on those responsible for laying out and managing our forests to design them with vision and forethought in the certainty that they will be occupied by deer, no matter how many miles of fences are erected. Better ways of managing red deer in forestry can be found than harassing them day and night, in season and out, shooting hinds in milk and stags in velvet. These things will go on because they are easy options, until public indignation brings them to an end.

10 • Sika – The Phantom Whistler

Imagine yourself in a deep pine wood just as the October dusk thickens. It is an eerie time; maybe an owl hoots; a time when the most hardened woods' lover starts to have that uncomfortable feeling that something is watching from behind.

All at once the silence is split by a shrill whistle, incredibly loud, first rising in pitch then falling away, repeated and repeated again, ending in a deep and savage-sounding grunt. Even if you are expecting it, the rutting whistle of a sika stag always makes the hair rise on the back of your neck. If you are unsuspecting, it is enough to make you jump out of your skin, if not run pell-mell to the safety of the car. No wonder that a white-faced walker once burst into the local pub near Wareham in Dorset and shouted 'There's a lion in the woods!'

If you can restrain your panic, stand like a statue, straining your eyes and ears in the gloom for the smallest indication of where the stag may be. He will be restless, moving about his chosen rutting stand and hoping, no doubt, that his piercing whistles have charmed a wandering hind into drifting his way, for that is the whole idea. He has probably

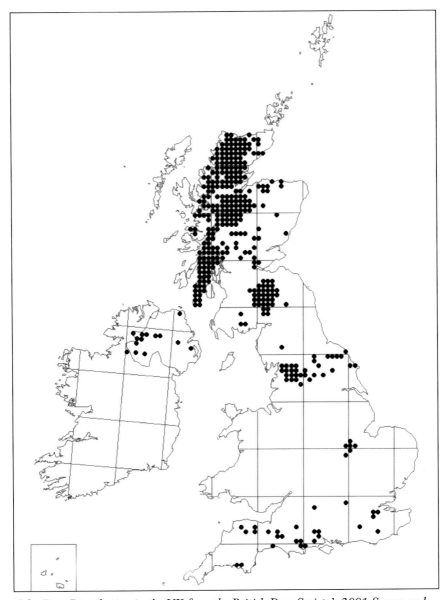

Sika Deer Distribution in the UK from the British Deer Society's 2001 Survey and reproduced by kind permission.

heard your own movement and may come to investigate. You hear the slightest whisper in the leaves or, more likely, catch a flick of white from the polished tips of his antlers from which you can locate a shadowy outline of the stag himself as he stands, deeply suspicious, among the sombre pines.

SIKA DEER (*Cervus nippon*)

Male: stag. Female: hind. Young: calf.

SUB-SPECIES
Most feral populations in the UK are Japanese sika (*Cervus nippon nippon*) which are described here. Some other sub-species or hybrids may be encountered. Manchurian sika (*Cervus nippon mantchuricus*) are larger, with spreading antlers, reddish in velvet. Formosan sika (*Cervus nippon taiouanus*) are slightly smaller than Manchurian; yellowish in summer coat rather than red. A few black sika in Whipsnade Park may be hybrid *Cervus nippon keramae*.

HYBRIDS WITH RED DEER
See text.

Japanese sika

LENGTH: 150cm (60in).

HEIGHT AT SHOULDER: 82–90cm (32–35m).

WEIGHT: Wide variation. Stag c 70kg (160lb); hind 35–40kg (80–90lb), live weight.

PELAGE: Summer: bright brown with creamy spots; white caudal disc, fanned out in alarm; black border and stripe above tail which is shorter than fallow, 10cm (4in), and generally white. Winter: hinds are greyish-brown, practically unspotted; stags blackish, with short bristly mane; both have prominent tail patch. Fawns: brown and spotted.

ANTLERS: Similar to red deer but typically eight points. Shed April. Clean early September. Velvet greyish-black or pinkish.

BREEDING: Rut September to early November. Stags whistle loudly. Calves born June; one only.

DISTRIBUTION: Widespread and expanding in Scotland. Locally abundant in England and Ireland. Herds in several parks.

HABITAT: Thick, often damp, coniferous and broadleaved woodland and wooded fringes of moorland; boggy scrub.

FOOD PREFERENCES: Mainly herbs and grass. Some browsing. Bark stripping often associated with disturbance.

HABITS: More nocturnal than red or fallow. May travel some distance between cover and feeding area. Small groups of stags, or hinds and juveniles. Old stags are very secretive. Young wander quite widely. Groups are slightly larger in spring. Likes a wet wallow. Some stags score the bark of large trees with their antlers.

VOICE: See text.

IDENTIFICATION: Distinguish from red by large white caudal disc and white tail. White gland below hock and V-shaped white mark on forehead. From fallow by round antlers, rounded ear, hinds greyish in winter. From roe by obvious tail, larger size, red-deer-type antlers.

Such an encounter can be breathtakingly exciting, a rare glimpse of a proud and beautiful creature whose way of life, even now, is not fully understood. If you are very lucky the stag may come out of the trees so that you can see his spirited bearing, his compact powerful body, and the short, strong eight-pointed antlers which are typical of his species. All too soon he will fade away once more into the gloom. Ten or fifteen minutes later another repeated triple whistle will show that he has forgotten his temporary alarm in the urgencies of the rut. If, on the other hand, he has circled to get a puff of your scent, he will reveal a danger to other deer nearby with a loud series of high-pitched whistling barks ending with a short, deep grunt of disapproval. Then you know that the vocal display is over for the time being and a move to another part of the wood is probably advisable.

Seven-point sika stag. Note the prominent rump patch and a longer tail than fallow. Sika also have a v-shaped white mark on the forehead and a white patch on the hind leg which can show below his hock.

A mature sika stag can be called during the rut by an imitation of his whistle. No doubt he thinks that another male has arrived to challenge his occupation of a prime rutting stand and he comes to see him off and do battle if necessary. Sika are determined fighters, quick and savage, and many stags will have broken antlers by the end of the season. Most people are unable to make the whistle loud enough with their lips alone. Some form of artificial aid is needed and finding something suitable needs a little enterprise and ingenuity. Eric Masters, the Forestry Commission's senior Ranger at Wareham was probably the first man to call sika in this country. He found that a squeaker from one of his children's rubber toys had just the right effect. Each year he used to go into the local toyshop and squeak all the rubber toys on display in order to make his choice for the calling season. The actual squeaker was all he wanted, the rubber duck, bear, dog, or whatever, being rejected. It caused a certain amount of consternation in the toyshop and I never learnt whether his children appreciated a regular supply of new but mute toys.

Sika are, in fact, the most vocal of all our deer. Steve Smith, who spent endless hours studying and photographing them in the New Forest, identified no fewer than nine different voices which the sika stag will use throughout the year. As well as the alarm bark and the whistle, he named them the *agonised groan*, the *wail, lip blowing, gargling, laughing wickers, bleating*, and the *terror scream*. This does not take into account all the different sounds which are made by hinds and calves. The fascinating thing about deer watching is the more you study them, the more complex and interesting their lives turn out to be. Just one species of deer, let alone the six we have, is more than enough for a lifetime of study and interest.

An oriental deer

Sika deer originate in the Far East. Different races or sub-species can be found from south-eastern Siberia, Manchuria and Korea, through China to Vietnam. They are also found on Taiwan (Formosa) and a number of the Japanese islands. In spite of obvious differences in size, appearance and voice, sika are very closely related to red deer and hybrids between the two species are fertile. It is even thought these days that some of the supposed sub-species of sika originating from the mainland of Asia may, in fact, be hybrids, the result of deliberate breeding experiments by the Chinese in the unrecorded past. This has only become important recently with the spread of feral sika in Britain which has led to hybridisation with our native red deer.

Sika of the typical Japanese race were first brought to this country in 1860, when a pair were given to the London Zoo and others were sent to Powerscourt in County Wicklow. By 1884 the Powerscourt herd had increased to one hundred and a number were given away or sold to other parks in Ireland, as well as to deer enthusiasts in England and Scotland. Very soon small colonies became established in the wild, either as a result of escapes from parks, or where they had been deliberately set free.

Before World War One specimens of other regional types of sika were imported, notably the Manchurian which are much larger in size than the Japanese race, standing 96–99cm (38in) high at the shoulder compared to 65cm (25½in). They have heavier, generally wide-spreading antlers, with reddish velvet instead of grey-black. Unfortunately, the late-Victorians' passion for experiment was not matched by their record keeping, and little is now known of the origin of some of our sika deer herds. It has, however, been supposed that sika colonies now established in Lancashire, central Kent, County Fermanagh and County Tyrone may have at least some Manchurian blood.

Like most successful animals, sika are fairly adaptable in their habitat requirements although they seem to do best on acid soils and in places where they can move freely between thick coniferous woodland and fields or moorland. They like very wet places. I remember watching four big stags walking across Morden Bog in Dorset and was very surprised when I found it was far too treacherous to follow them. It is curious how this Dorset colony, although numerous, restricts its range to the acid sands of Wareham Heath, without expanding on to the surrounding dry chalk hills where one only finds the occasional straggler. It owes its origin to an introduction of Japanese sika to Brownsea Island in Poole Harbour, about 1896. The story is that some of them swam ashore the very same night to found the colony which still exists on the Purbeck side of Poole Harbour, while others came from a park at Hyde, north of Wareham, from which they escaped soon after World War One. In the pre-war days Wareham Heath was still a big expanse of heather and bog with clumps of scrub and rhododendron. When the Forestry Commission started large-scale planting in the 1940s the sika obviously welcomed the extra cover available to them, and have flourished there and in the surrounding area ever since.

Not far away there is another colony in the southern part of the New Forest, originating from the Beaulieu estate. As usual, nothing seems to be known of its origin, but the antlers grown by New Forest stags are quite distinct in type from the Dorset sika. There is another important sika population on the Yorkshire-Lancashire border in the neighbourhood of Bolton-by-Bowland. These deer were originally released about 1907 by Lord Ribblesdale. In Scotland, sika have flourished following a number of introductions late in the last century. Until fairly recently the main areas were Argyll, the south side of Loch Ness, and central Sutherland, but probably as a result of the afforestation programme they have been spreading and increasing in numbers. Sika can now be found more or less continuously on forested ground up the west coast from Kintyre to Sutherland. They have also extended north-eastwards to a line roughly from Loch Fyne, towards Elgin where the great forests of Moray and Speyside lie open to colonisation. If that is good news for the deer watcher, it is ominous for the professional forester faced with the damage which these deer can undoubtedly do.

A similar threat is posed by sika which originate from Dawyck Estate,

Sika hind in summer coat. Note the wide black band surrounding the rump patch which distinguishes them from fallow.

Peebles. After many years of stability they have now pushed out into the Border hills and are likely to colonise much or all of the enormous area from Galloway to Kielder and beyond, which is now growing up into forest. In Ireland, sika are firmly established in three main areas: County Kerry, County Wicklow and County Tyrone, with other less numerous colonies in County Dublin and County Fermanagh.

Points for identification

A number of parks in Britain hold sika or have held them at one time. Outliers or small breeding communities are quite likely to be found somewhere in the vicinity. The difficulty in mapping the distribution of sika is that identifying them can cause confusion. In size they are similar to or rather smaller than fallow. When alarmed the white hairs of their caudal patch can be puffed out in the same way as roe, while the stags have rounded antlers like red deer. Females (*hinds*) and young (*calves*) need to be studied very carefully to avoid making errors in identification. While a mature stag, especially in his dark winter coat, is readily identified, more than one juvenile sika has been shamefully taken in mistake for a roebuck. Just to make the confusion worse, in some districts the dark type of fallow are referred to as sika or 'Japanese deer'.

The points to look for are illustrated in the drawing on page 133 primarily the V-shaped white mark on the forehead; a fairly short rounded ear; the tail shorter than fallow but longer than red deer and predominantly white in colour. Perhaps most indicative of all, if the animal is standing in the open, is a white tuft of hair on the hind leg just below the hock.

In summer sika have an attractive pattern of well-defined white spots on a dark chestnut background with black markings. Before the rut in the autumn the older stags grow a shaggy, nearly black coat with a distinct mane. Soon after, the hinds follow, changing into a thick silver-grey coat which gives them a ghostly look in the half light. Like most young deer, the calves are born spotted, changing rather early into a winter coat similar to the adults.

Like red and fallow, sika are herding deer and through most of the year one can expect to see parties of hinds and calves feeding in company, but each species has its individual quirks. While a strong family bond is obvious among red deer, sika calves go their separate ways before the arrival of the next generation. Stag calves wander off to join their seniors in the spring while even the mother-daughter bond is broken quite soon afterwards although the way this is done is not really understood.

Even the habits of the stags are something of a mystery. Once the rut is over, little will be seen of them. One can assume that they wander off like red stags in small bachelor parties in search of better food or peace and quiet. They are, incidentally, the first colonisers of new ground. In the thick woodland beloved of sika, quite a number of cautious old stags could escape notice completely by lying low and only emerging to feed very late at night. It only shows how much there is still to learn about our deer and what marvellous opportunities exist for anyone with the time and the enthusiasm to study them.

Damage and disturbance

Another real puzzle is the food they eat and the damage they do. For years we considered the sika to be primarily a grazing animal and therefore a potential nuisance to farmers, but doing less harm to forestry than any of the others. The exception to this rule were the New Forest sika which appeared, for whatever reason, to take quite a lot of woody browse and to do a significant if not serious amount of damage to trees because of the stags' habit of prodding them with their antlers. Nobody up to now has ever suggested why there should be this anomaly. Reports of serious bark stripping damage then began to trickle in, first from abroad and then from Scotland. This was quickly linked to changes is amount of disturbance which the deer had to suffer. Like many questions of deer behaviour, it is common sense when you think about it. We are dealing with an adaptable animal which is successful in a variety of environments. Obviously the sika change their behaviour and feeding strategies when their normal preference for feeding at dawn and dawn in the fields is threatened. When one considers the conditions for deer in the New Forest, subject to constant harassment by people and dogs, it is no longer surprising that they rely more on thick wood-land to support them than do their cousins in the tranquil woods and small fields of Dorset, or the remote forests of Inverness-shire.

What sort of disturbance is likely to be involved in this change of habits? One reason is self-evident. We live in an age of leisure and increased mobility. This has resulted in a vast increase in the recre-ational use of woodland. These days, the woods are full of ramblers and bird watchers, dog walkers and picnickers, joggers and orienteers. Near any centre of population almost every large stretch of woodland is subject to human incursions from before first light until after dark. To this one has to add normal forest operations which have become more frequent with the intensity of management, and periodic disturbance even of the smallest copses because of pheasant shooting which might otherwise have served as sanctuaries. No wonder under this stress the deer have become less willing to trust themselves to the danger of the open fields. Failing a leisurely feed of grass or cereal, in hunger or frustration they start to strip the bark from the trees.

There is another more subtle influence on deer movement. For what-ever reason, deer detest sheep. Maybe it is their smell, the noise they make or the presence of the shepherd and his dog. Certainly the fences which are erected when a farmer changes from cattle to sheep make significant barriers for deer movement, although they could if pressed jump them without difficulty. Deer always prefer to creep rather than jump. One owner in Scotland queried why the sika were doing serious

bark stripping damage in one of two similar and almost adjacent woods but not the other. The answer indeed was sheep – the plantation where damage occurred was surrounded by fields full of them but cattle were allowed to graze round the other. Damage is also necessarily a side effect of density. In the north where the majority of sika are to be found, the forests which they have colonised were never laid out with deer control in mind. Where they cannot expand into new country, breeding populations are bound to rise to the point where both farm and forest damage becomes unacceptably high.

Hybrids with red deer

The fact that sika sometimes hybridise with red deer has already been mentioned. Because the two species are fairly closely related, hybrids are themselves fertile. Where the two species of deer exist side by side there must be concern that a mongrel race will develop. Many species are very widely distributed over the world and through isolation and adaptation to different conditions over thousands of years have developed varying characteristics which distinguish one local race from another. These variations must be sufficiently fixed and definitive to merit classification of related types either as sub-species or even as distinct species. When man in his interfering way brings two divergent

A sika whistling in the rut. Hearing this late at night can be quite alarming!

races together once more, for example in a deer park, it is surprising how little notice one may take of the other. There are, no doubt, subtle differences in behaviour, particularly in breeding behaviour, as well as voice, and possibly smell, which combine into a code recognised by near relations. Others may be genetically very similar but are sufficiently remote in ancestry to have developed a different behavioural language so that the response is not triggered.

For example, one park which I had the privilege of managing for many years is stocked with fallow, red deer and sika. During the autumn, the air is lively with the different calls from rutting males of all three species, but although the red and the sika could undoubtedly hybridise, no authentic cross has been recorded over the century since sika were first introduced. They take no notice of one another, although a sika stag may be established under one oak tree while a big red stag bellows his head off fifty yards away under the next. Despite this, hybridism has undoubtedly occurred in the wild, and deer of mixed race can now be seen in the Wicklow Mountains of Ireland, to the south of the Lake District in north-west England, and have been reported lately in western Argyll.

Sika-red hybrids have been taken to be red deer with aberrant coat colour. As long ago as January 1961 a yearling red stag with a most unusual black and white tail patch was observed in north Lancashire running with two normal-coloured youngsters. As a three-year-old his antlers were those of a typical three-year-old six-pointer red deer. However, more deer of the same coloration started to be noticed in the same area. Five years later a stag was noted in the same area as 'a heavy, stocky, dark-coated, fully mature beast with a short face and an unimpressive eight-point head. He had a pure white rump patch bordered by black and with a black line down the tail.' As time went on successive generations of these undoubted hybrids became more sika-like as if the original outcross with red deer had not been repeated. Dr Delap, an acute observer of the deer of north-east England, described these deer as:

Two young sika stags. Juvenile antlers may resemble those of roe and lead to mis-identification. Photo Brian Phipps

. . . red up front, and sika behind. In winter the tail patch is most startling, and all display the light hock gland of their oriental parent. As with pure-bred sika, the black line on the tail is variable, tending to be absent in the young. Some have notably grey-brown coats and a few the rather rounded ears. In summer most of the animals resemble slightly -stunted red deer although one hind had a fully spotted coat and one young stag a scattering. The calves have a rather light coat and like all the young-sters look frailer and smaller than pure red deer of comparable age, the stocky build coming with maturity.

How can one imagine the original red-sika cross occurring in the wild when in the close conditions of a park there seems to be so little inter-action between the species? There are some interesting pointers, and some other areas where the deer watcher can use his developing under-standing of deer mentality to visualise the likeliest circumstances in which hybridisation might take place.

It has already been mentioned that some of the mainland sub-species of sika may, in fact, be a result of animal breeding experiments between sika and red deer by the Chinese long ago. If this is so, one should be able to produce an animal similar in type to, say, a Manchurian sika, by deliberately crossing a Japanese sika stag with a red hind and then allowing the progeny to breed among themselves to stabilise the type. This has indeed been done and the result has been claimed to be an animal more or less indistinguishable from Manchurian sika.

The next point to look out for would be that if the Manchurian actu-ally are the result of a cross with red deer, one would expect less of a behavioural barrier between them and red deer than between red deer and the pure Japanese race. A third possibility might be that the sika that were released or escaped from parks were already hybrids, the result of more recent experiments. In spite of the lack of records kept by park owners, there are some fascinating pointers which one can extract even from quite easily available sources. We know, for example, that Walter Winans had Manchurian sika in his park at Surrenden Dering up to World War One and that the sika surviving in the wild in central Kent are the descendants of escapees from this park. However, in this area there are no red deer. In north Lancashire, in contrast, red deer have existed for centuries and although the original sika introduced for hunting were supposed to be of the Japanese sub-species, Mr J. Robinson talking to the East Anglian branch of the British Deer Society in 1972 said: 'It is generally considered that the Japanese sub-species

Sika hind and calf. Photo Brian Phipps

which was originally introduced proved too slow for the hounds and was subsequently replaced by Manchurians.' Dr Delap, whose name has already been mentioned, supported this point of view and noted that the stags' velvet is a startling red and black colour whereas the Japanese sika has grey or black velvet. Another curiosity of this sika population is that family parties of a hind, calf and follower (calf of the previous year) are frequently seen, as one does with red deer, while the undoubtedly pure Japanese sika in Dorset drive off their yearlings before the birth of the next calf.

The next possibility is that the deer originally released or escaping were themselves hybrids. In *British Deer and their Horns* (1897) J. G. Millais includes a drawing of the skull of a seven-point hybrid Japanese and red deer stag bred at Powerscourt, County Wicklow, 1894. Sika from Powerscourt were sent, admittedly before this date, to Colebrooke Park, County Fermanagh, which already contained red deer. On at least two occasions hybrids between the two species were noticed, the first, according to G. K. Whitehead, occurring in 1885, and another two years later. Whitehead also states that there is a possibility that Manchurian sika may also have been brought in. Could it be that with the passion for 'improvement' of those days the sika sent from Powerscourt had already been crossed with red deer and would therefore have no inhibitions about doing so again? The Lord Powerscourt of those days was a prominent member of the Acclimatisation Society, a body which was dedicated to the introduction and crossing of many exotic species of animals. It may be significant that hybridism is widespread now among the deer of Wicklow and Fermanagh, while in another part of Ireland, County Kerry, the species have not crossed although they live in the same area.

There is no suggestion that the sika originally taken to Argyllshire were anything but pure Japanese. How is it then that hybrids have now been reported from that area as well? Although sika were introduced to Kintyre as long ago as 1893, and have been extending their range in red deer country ever since, hybrids with red deer have only recently been reported. In that long time the population has twice risen to a peak when they became a pest and rigorous control measures were imposed, once about 1925 and again during the war. It is not therefore purely population pressure, nor the fact that colonising sika stray more and more on to red deer ground and therefore come into contact with them. This long period of colonisation without hybrids bears out the supposition that the original stock were of pure Japanese race. What other factor could have influenced a change in such a long established pattern of behaviour? It may well be forestry fences and the treatment of deer inside them.

The current method of protecting new plantations from deer damage is to fence them. Deer of all species will attempt to get over or through these fences because the area of ground may lie within their traditional range, because it is important to them for food and shelter during the winter, or merely because the young trees, often with the benefit of fertiliser having been applied, are attractive to them as browse. Every attempt will be made, especially in the early stages, to eliminate any deer which succeed in crossing the fence, but as the trees get taller these deer are more difficult to find and a total wipe-out becomes progressively less likely. If one assumes that in one particular plantation there happened to be a few red hinds and a male sika left alive during the rut, hybridisation will almost certainly result in the total absence of herd structure for either species. The same thing could happen in any area of the country where deer herds are broken up by harassment to the point where females in rut are deprived of the normal outlet for their sexual appetites.

Apart from the interest to the deer watcher of identifying a sika hybrid as something out of the ordinary, one could ask whether it matters if deer of intermediate type between two beautiful animals should occur in different parts of the country. Personally, I feel that the integrity of our deer species should be preserved and that rigorous steps should be taken not only to prevent the spread of hybridisation, but to get rid of the existing stocks which are undoubtedly a threat to red deer. The danger of losing the original strains of domesticated animals and plants has only just been recognised. Commercial hybridisation may serve a short-term aim possibly by increasing yields of milk, meat, eggs, or grain, but the original genetic material must at the same time be preserved for the future. Our native deer deserve no less.

11 • *Chinese Water Deer – It's a Teddy Bear!*

Between our smaller exotic deer, the cuddly-looking Chinese water deer usually has to act second fiddle to the star turn of the successful muntjac. The species deserves better than this, for water deer are interesting and charming animals which have made more progress recently in establishing themselves in the wild than most people give them credit for. Imported by the Duke of Bedford in the early part of the twentieth century, they became well established in Woburn Park so that by 1929 some were sent to Whipsnade Zoo where a numerous colony built up. The inevitable escapes from these two centres were the main starting points for the feral population we now possess.

Chinese water deer are often dismissed as scarce and local as well as being not really hardy in our climate. The British Deer Society Survey map soon confounds this idea. From their point of origin in Bedfordshire they have spread into the fens towards the Wash and Lincolnshire. Escapes from Whipsnade may have contributed to the occupation of North Bucks, and there is a solid block of the

Norfolk/Suffolk coastal strip where they have become established. Water deer did get the biggest belly-laugh of all out of the scientific community by establishing themselves in Woodwalton Fen, a nature reserve of international interest and significance, and building up a numerous colony before anyone realised that they were not muntjac. Since then, this community has been the subject of a fascinating study which is all the more significant because the conditions there closely resemble the natural habitat of the species in China and Korea.

Woodwalton is one of the last places where one can get an idea of the marshy fenland wilderness which once extended from Bedford to the Wash. Inside the reserve deep drains were dug so that peat and hay could be cut and carted easily by boat. Grass and scrub woodland on the drier parts give way to forests of reed where the water table is higher, and the whole fen is subject to winter flooding. Across the boundary dyke is nothing but flat, rich reclaimed land as far as the eye can see. In this primeval swamp the water deer have reverted to a natural pattern

Chinese water deer buck. This species has no antlers, but mature males have long tusks, which can be seen here. Photo Brian Phipps

Buck in winter coat.
Photo Brian Phipps

of behaviour. This has become modified in the artificial parklike conditions of Whipsnade Zoo where numbers of them can he seen together on the outer fields. The fenland deer are much more solitary, only congregating regularly in the spring when shortage of food or flooding of their territories forces them to feed in the fields.

There should not really be any difficulty in distinguishing between the hunched, piglike silhouette of the muntjac and the more upright stance of the water deer. When they look at you they have large furry ears with dark button eyes and nose making them resemble a cuddly toy in some ways. Like muntjac, they are very primitive in ancestry, to the extent that neither sex grows any antlers. The buck, however, has long curved tusks which in adult animals jut down well below the jawline. These tusks have been the subject of an interesting study. Although they are difficult to extract they have a limited fore and aft movement which is controlled by attachments to certain muscles of the face. When the buck is feeding or in repose the teeth lie back towards the angle of the mouth where they do not interfere with chewing. However, if the animal is startled or frightened the facial muscles contract in a form of snarl and the teeth hinge forward making them more easily used as weapons. Despite this, water deer appear to be very inoffensive animals and they do not have the reputation for self defence earned by the muntjac. They do fight among themselves. During their rut in December one can find chunks of hair lying about where two territorial bucks have come to blows. As well as the loss of hair the ears take rather a battering. Serious injury is very unlikely.

Apart from the rut there is a period of aggression and chasing in March to May before the young are due to be born, but outside these times water

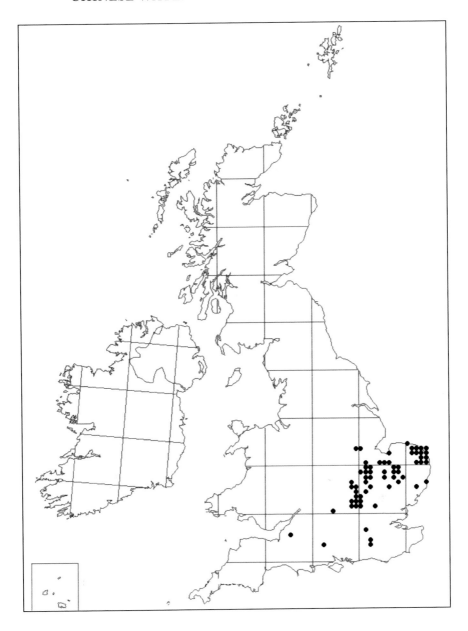

Chinese Water Deer Distribution in the UK. From the British Deer Society's 2001 Survey and reproduced by kind permission.

deer are rather solitary. The territorial system seems more flexible than, for example, with roe deer. As could be expected, territories are marked by scrapes and scent. Lacking antlers, the buck uses his tusks to polish

CHINESE WATER DEER (*Hydropotes inermis*)

Male: buck. Female: doe. Young: fawn.

HEIGHT AT SHOULDER:	50cm (20in).
WEIGHT:	Buck 12–19kg (26–42lb); doe 15.5–17kg (34–37½lb), live weight.
PELAGE:	Summer: reddish-brown, no spots. Winter: pale to dull brown flecked with grey. Coarse hair. Fawn: mostly dark brown; two rows of yellowish-white spots which soon fade. Tail: 6cm (2in) not usually visible.
ANTLERS:	None.
BREEDING:	Rut December. Fawns born June–July. Multiple births, up to six, average two to three. Young mature at six months.
DISTRIBUTION:	Bedfordshire, Northamptonshire, Buckinghamshire, Norfolk and adjacent counties, Shropshire. Local colonies elsewhere due to escapes.
HABITAT:	In China, swamps and reed beds; riverine grasslands. UK: agricultural land and woodland; reed beds and scrub.
FOOD PREFERENCES:	Primarily grazers. Growing shoots of grasses, sedges etc. Some herbs, browse in winter. Potatoes, carrot tops and roots.
HABITS:	Rather solitary. Males territorial, especially in rut. Adults static in home range unless disturbed. Peak activity around sunset, but active day and night. Territory marked by scent from preorbital gland and by urination in scrapes.
VOICE:	High-pitched alarm bark. Frightened deer scream. Clicking and squeaking noises mostly during rut.
IDENTIFICATION:	Smaller than roe, longer legs than muntjac. Mature bucks have long, 7cm (2¾in), curved movable tusks visible in upper jaw. The female's tusks are smaller. Buck's tusks longer than muntjac. Prominent hairy ears, rough coat and 'button' nose. Hind legs longer than fore. Short dark, hardly visible, tail; no white tail patch.

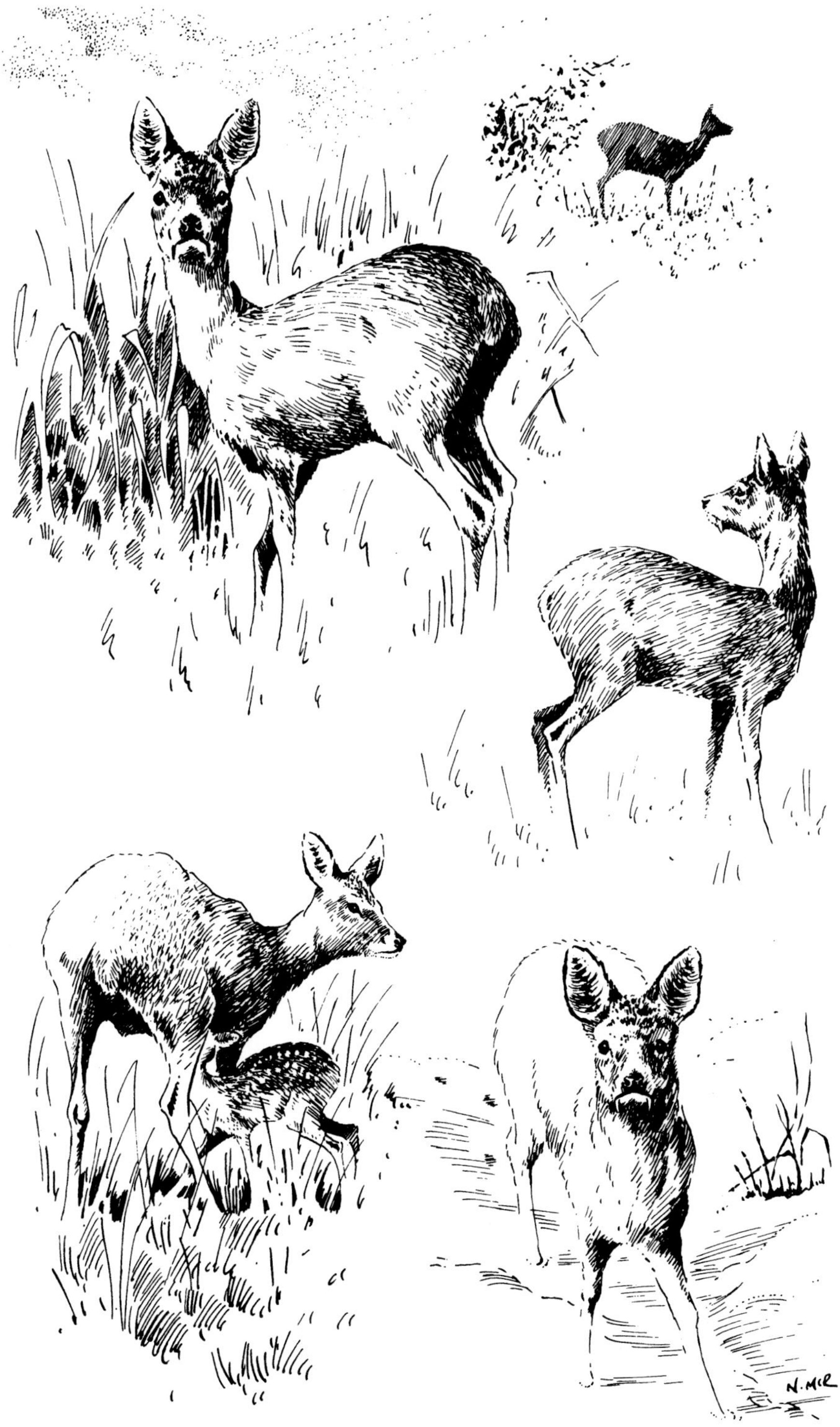
N. McR

Chinese water deer doe. Note the powerful hindquarters.

small sterns which are then anointed with scent from the pre-orbital gland. Water deer also bark, probably in challenge as well as alarm. Curiously enough, the peak period for barking at Woodwalton Fen is from June to August rather than in the December rut.

Heavy mortality has been reported from time to time among Chinese water deer in bad weather. As they are native to northern China and Korea, where the climate is comparatively cold in winter, these losses may be due less to hard weather than to lack of food or the attacks of parasites. Their winter coat is thick and harsh which should be a good protection. In deep snow their movements may be hampered. This makes them vulnerable to predation by foxes which owing to their light weight can run on snow without breaking through the crust.

The water deer compensates for this liability to heavy natural mortality by having the highest potential reproductive rate of any of our resident deer. Twins and triplets are common while litters of up to six have been recorded. The likelihood of survival for a big litter is limited, besides anything else by the fact that the water deer doe has only four teats so that weaker fawns would be likely to be ousted at meal times.

Even so, with an average litter size of nearly three the water deer is capable of rapid recruitment after a crash.

In natural conditions, water deer appear to be primarily grazing animals preferring grass, sedges and herbs as well as a certain amount of browse such as the tips of bramble shoots, willow and so on. Although their presence on the fields in early spring may worry the owner of the crops, they have not so far been shown to do appreciable damage and one suspects that they have been blamed at one time or another for a certain amount of harm which was actually done by rabbits. A deer out in a field is not necessarily eating the crop, and only close observation will produce an answer. They do nibble roots such as potatoes and carrots and might not be at all welcome in a market garden or nursery.

For anyone living in the Midlands or east of England the opportunities for studying Chinese water deer are good. They may be observed at all times of the day in the grassy expanses of Woburn Park or at Whipsnade Zoo. Getting to know them in the wild depends on local opportunities and permission. The time when they are most active is the hour round and after sunset. Having selected a promising area, possibly the best way of locating a colony is to spend a little time in late winter or early spring travelling the roads with a pair of binoculars, trusting to catch them out on the farm crops. An approach to the owner of the nearest woodland for permission to watch them from March until mid-summer would not conflict with any game shooting and, given confidence in your bona fides, permission may well be forthcoming. At least when you see a Chinese water deer for the first time you will not be caught like the small boy at one of my lectures who, seeing a slide of Chinese water deer, called out, 'Mummy, it's a teddy bear!'

12 • *Where did they Come From?*

Fossil deer

About fifty million years ago, creatures existed which can be seen to foreshadow modern deer. The first which can be recognised as possibly being related to present-day forms date from about thirty-five million years ago. Very early types had bony outgrowths from the skull. In deposits of that era can be found the remains of creatures like *Syndyoceras*, a distant relative of the camel which had one pair of divergent horns on the nose and another pair at the top of the skull. The probability is that these horns were fur-covered like those of the giraffes of today. Some fossils have been discovered with horns showing a number of points. Although they closely resemble antlers they were, in fact, merely bony outgrowths which were never shed like true antlers. Some authorities have even suggested that one group of animals, the *Merycodonts,* from North America, had antler-like structures with points which were bare bone but non-deciduous (not shed annually). Certainly some of the skulls would pass for roebuck with their short backward-pointing antlers and marked coronets. They are supposed, however, not to be in the evolutionary stream which led to the deer but may be related to the American pronghorn.

The world's earliest known deer, *Dicrocerus* and *Stephanocemas*, have been discovered in Miocene deposits of Europe and Asia dating back fifteen to thirty-five million years. They had long pedicles and forked antlers which were undoubtedly shed periodically because cast specimens have been found. These were small, muntjac-like deer with prominent tusks in the upper jaw. It may be a reflection on the primary value of antlers that the upper canine, which is long and sharp enough to form a useful weapon in these early forms, and also in modern muntjac and Chinese water deer, has disappeared altogether as the structure and size of antlers has developed. These two can therefore be regarded as a very primitive deer, almost living fossils, which have survived virtually unchanged since Miocene times. Such immense lapses of time are difficult to comprehend until one is told that there were animals resembling muntjac in Europe before movement of the Indian sub-continent pushed up the Himalayas!

The other significant development was a reduction in the number of bones in the foot, presumably in response to the need for escape from predators by rapid flight. The 'odd-toed' ungulates which culminated in

the horse, run as it were on the tip of one finger, while the 'even-toed' ungulates (animals with cloven hooves) including cattle and deer have developed digits three and four equally. The dewclaws are more or less vestigial remnants of digits two and five.

The Ice Age

Ice ages are now known to have occurred periodically throughout the earth's history. The present one in which we still live is supposed to have begun nearly two million years ago. In geological terms it has been a period of rapid climatic fluctuations which led to a speeding up of evolution. Animals had to adapt rapidly or die. As the climate turned colder and ice sheets began to spread from the poles and mountain chains, immense quantities of water became locked up resulting in a drop in sea level. One effect of this was to expose various land bridges which allowed pedestrian species to colonise new islands and continents. There was, for example, a wide expanse of land between Asia and America, where the Bering Strait now lies. The horse, which evolved in America, crossed into Asia, while the bovines and some primitive mastodons passed in the opposite direction. The deer are thought to have evolved in Eurasia, but the present New World forms related to the white-tail arrived in America as long ago as the Pliocene. The moose, the reindeer, and the wapiti, a close relative of our own red deer, are more recent arrivals from Asia. With the virtual drying up of the North Sea, Britain was open to colonisation from European forms. There was also for a short time a land bridge to Ireland.

It is a mistake to think of the Ice Age as 1,700,000 years of continuous winter. During the Pleistocene epoch there have been continual fluctuations of temperature with long periods, the *Interglacials,* when the climate was considerably warmer than at present, and when the sea level must once again have risen.

The incredible giant deer

At the beginning of the period, deer which were very similar to modern forms began to emerge. In addition, there were some spectacular monsters which have sadly died out. The greatest of these was the giant deer, *Megaloceros,* which stood nearly 2m (6ft) high at the shoulder, roughly the same as a Canadian moose. It had antlers palmated like a

The muntjac has possibly survived virtually unchanged since Miocene times, fifteen to thirty million years ago!

fallow deer, weighing in some specimens more than 40kg (90lb) and measuring more than 3m (9ft) from tip to tip.

This magnificent animal seems to have been highly successful, inhabiting a large area of what is now Europe and the broad grassy plains which developed between England and Europe and also on the Continental shelf as the Atlantic retreated. It must have needed an abundant food supply to grow these enormous antlers which, in common with every deer species, were shed and re-grown every year.

Even so, giant deer seem to have coped well with colder periods when the country must have been a form of tundra with birch, willow and juniper predominating. Their remains have been found in association with bones of reindeer and also such cold-adapted species as the woolly rhinoceros, cave lion, bear and mammoth. In more genial times they existed with the hyena and the hippo. Many complete skeletons have been found in Ireland, normally lying below peat deposits in shell marl. The theory is that these deer were drowned after getting bogged, or by breaking through the ice while attempting to cross lakes in the winter. In England most of the remains of *Megaloceros* have been broken up in floods or chewed by large predators such as the hyena. It has been

suggested that giant deer succeeded in crossing the land bridge to Ireland while some of their major predators failed to do so. Certainly they must have been very numerous there at one time.

Did our remote ancestors actually pursue a giant deer, with its huge and daunting antlers? The evidence is conflicting, mostly depending on cave drawings by Palaeolithic man. Most authorities claim that it became extinct in Britain about ten thousand years ago, but possibly survived even as late as two thousand five hundred years ago in the region of the Black Sea, although no fossils have been found dating later than ten thousand years Before Present (BP). One can see two complete reconstructed skeletons in the British Museum (Natural History) in South Kensington, London, which give a dramatic impression of this formidable animal. Sets of antlers may also be found in a number of stately homes open to the public. There is, for example, a notable pair in Longleat House near Bath.

Modern deer

True red deer and roe appear in Interglacial deposits of more than half a million years ago. They are likely to have died out and recolonised the country after the peak of each glaciation. Red deer seem to have become smaller, but the roe were identical in appearance to the deer that steal our roses today.

The story of the fallow deer is neither so continuous nor so well documented. Remains of deer resembling modern fallow, usually called the Clacton fallow deer, have been found in great numbers in late Middle Pleistocene deposits at Swanscombe in Kent, and earlier types are also recognised.

Fallow deer are generally believed to have died out in Britain at the time of the last glaciation. Recolonisation as the climate improved is doubtful. However, the species did survive, probably in countries to the east of the Mediterranean sufficiently unaffected by ice and snow. Evidence about their fortunes in this period is fragmentary but the absence of fallow remains in Mesolithic and Neolithic sites in Britain suggests that fallow did not succeed in spreading northwards again before the rise in sea level cut the land bridge from Europe about eight thousand years ago.

Europe in Post-glacial times

Experts will argue that the Ice Age is not over, and that in geological terms we are merely enjoying a brief improvement in the climate accelerated or not by global warming, the result of human action. The last ten thousand years are usually termed Post-glacial, during which the ice sheet covering Scandinavia and northern Britain melted and the land was gradually recolonised. An alpine flora was followed by birch and later by pine, alder, aspen, hazel, oak, maple and elm. About seven thousand years ago most of Britain and Scandinavia was covered with predominantly broadleaved trees, with pine forests extending beyond the Arctic Circle. Since then the climate turned cooler, and the vegetation belts once more moved southwards.

With the melting of the ice sheet the sea level rose. At the same time the removal of the immense weight of ice from northern Britain caused southern England gradually to sink as the 'see-saw' effect was reversed. (This movement continues even now: London is sinking at a rate of nearly 30cm (1ft) every hundred years.) The combined effect was to form what is now the North Sea. The Straits of Dover were flooded about eight thousand years ago. This allowed only four to five thousand years for immigration to Britain after the ice retreated and the land became once more inhabitable. Our fauna is consequently impoverished in comparison to that of the Continent, but even so we received some notable colonists including the brown hear, lynx, wolf, the aurochs or ancient wild ox of the European forests, the reindeer, the elk or moose and the beaver.

Man soon followed and succeeded with his superior brain in gradually eliminating all these species over the centuries. The fallow deer and giant deer had both disappeared in the last glaciation, probably not to reappear. Elk and reindeer persisted into historical times. Destruction of large tracts of forest has been blamed for the certain loss of the elk by the ninth century, if indeed it survived that long, while reindeer lingered on in northern Scotland another two hundred years, leaving only red deer and roe to represent the Cervidae. These owe their survival in the intervening years to adaptability. They were able to take full advantage of what habitat was available even in a time of changing climate. Because of their evolution as prey species, they were well able to evade their enemies by concealment or flight and to make up losses by high reproductive success.

The impact of Stone-age man

There is plenty of evidence that in the densely-forested countryside of post Ice Age Britain red deer in particular were abundant and of great size. Even in Scotland antlers are dug out of the peat that dwarf the best present-day royal and show how much kinder the climate, and therefore the feeding, was for these animals in the dawn of history. In England the most dramatic demonstration of the way man started to use animals not just as food but to provide tools can be seen in the Neolithic flint mines in Norfolk called Grimes Graves.

For hundreds of thousands of years man used stones as weapons and tools, chipping them to make sharp edges and working them eventually into remarkably sophisticated shapes. Flint was ideal for this purpose, flaking away when skilfully hammered to produce a durable, sharp cutting edge. Flints lying on the surface would have been used at first, but long ago it was discovered that the best flint for the purpose, now called *floorstone*, had to be dug out of the chalk.

Flint mines have been found in Sussex which date back more than five thousand years, but work at Grimes Graves did not begin until several hundred years later. The main period of activity was between four thousand and three thousand eight hundred years before the present. One of the shafts has been opened up for the public to explore and is well worth a visit.

In the middle of the vast modern conifer plantations of Thetford Chase, Grimes Graves has been left as a clearing. It looks like a long-abandoned bombing range with an immense number of shallow grass-grown craters almost overlapping one another. In fact, this is a Neolithic industrial wasteland, each crater representing one of the seven

A reconstructed flint axe and red deer antlers used as picks in the flint mines at Grimes Graves in Norfolk.

or eight *hundred* mineshafts that were dug in an area of only about 40 hectares (100 acres). The deepest shafts are between 4 and 8m (12 and 25ft) in diameter at the surface and up to 14m (45ft) deep. They were dug to get at the vein of floorstone which can still be seen as a shiny black layer, perhaps 20cm (10in) thick, in the claustrophobic galleries at the foot of each shaft. Picks were made from red deer antlers by cutting off all the points except the brow and shortening the tool to a convenient length. They were used not only to loosen the overlying soil and chalk but to lever out the blocks of floorstone flint which were then carried up in leather sacks or baskets to the surface.

It has been estimated that between a hundred and a hundred and fifty antler picks were worn out in digging each shaft. As soon as they were blunt they were discarded, and can still be found in the rubble. About ninety per cent of the Grimes Graves picks were made from shed antlers which would have been picked up in the forest. Only ten per cent were from animals that had been killed. It is interesting to find that our ancestors were mostly right-handed because about two-thirds of the antlers

159

come from the left side of the head, which would have made the best shaped tool for a right-handed man.

By this calculation the digging of eight hundred shafts would have needed altogether between eighty thousand and a hundred and twenty thousand antlers over the period that the shafts were made, most of which remain to be found and examined as other shafts are excavated. It gives some idea of the abundance of red deer, some of which were obviously of great size.

As the wildwood was gradually cleared, primitive agriculture began to supplant the search for a bare existence by hunting. Predators like bears and wolves which previously kept the deer in check were relentlessly pursued as the enemies of domestic stock. Gradually the chase of wild beasts became the province of the king and his nobles, eventually to be ritualised into sport. It was this alone that protected them from total extinction through the Middle Ages.

The fortunes and numbers of British wild deer were never lower than around 1800, yet after less than two hundred years there are probably more now than in the lush days of the Interglacials. Nowadays we not only have our native red and roe deer, but fallow, sika, muntjac and Chinese water deer. The story of how this has come about, the early reappearance of fallow, the reintroduction of roe to southern England, of reindeer to Scotland, and the arrival the other exotic species to swell our native fauna has been told.

The repercussions of these releases are profound and to some extent worrying. In the meantime we have to live with nature and the fruits of our folly. At least there has never been a better time to go out deer watching!

Further Reading

Brown, M. B. *Richmond Park* (Robert Hale 1985)

Carne, P. *Deer of Britain & Ireland* (Swan Hill Press 2000)

Chapman, N. *Deer* (Whittet Books 1991)

Chapman, D. & N. *Fallow Deer* (Coch-y-Bonddu 1997)

Clutton-Brock, T. H., Guinness, F. E., & Albon, S. D. *Red Deer* (OUP 1982)

Goulding, M. *Wild Boar in Britain* (Whittet Books 2003)

Hawkins, Desmond *Cranborne Chase* (Gollancz 1980)

Hinde, T. *Forests of Britain* (Gollancz, 1985)

Hingston, F. *Deer Parks & Deer of Gt Britain* (Sporting & Leisure Press 1988)

Prior, R. *The Roe Deer – Conservation of a Native Species* (Swan Hill Press, 1995)

Prior, R. *Trees & Deer* (Swan Hill Press 1994)

Prior, R. *Roe Deer – Management & Stalking* (Swan Hill Press 2000)

Putman, R. *The Natural History of Deer* (Helm 1988)

Rose, R. *Working with Nature* (Grayling 2000)

Rouse, A. *Photographing Animals in the Wild* (Fountain 1999)

Sibley, P. *Ferney Wood – The Story of a Fallow Deer* (Whittles 2005)

Smith-Jones, C. *Muntjac – Managing an Alien Species* (Coch-y-Bonddu 2004)

Soper, E. A. *Muntjac* (Longman 1969)

Whitehead, G. K. *Deer of the World* (Constable 1972)

Yalden, D. *The History of British Mammals* (Poyser 1999)

Booklets
B D S – Mammal Society Booklets

Muntjac (Norma Chapman & Stephen Harris) 1996

Roe Deer (John K. Fawcett) 2006

Chinese Water Deer (A. Cooke & L. Farrell) 1998

Sika Deer (Rory Putman) 2002

Fallow Deer (J. Langbein & N. Chapman) 2003

A Field Guide to Deer in Britain (J.A. Lawton, Deer Study Centre) 2001

Index